Statistics

Other titles in this series

Linear Algebra
R B J T Allenby

Mathematical Modelling
J Berry and K Houston

Discrete Mathematics
A Chetwynd and P Diggle

Particle Mechanics
C Collinson and T Roper

Ordinary Differential Equations
W Cox

Vectors in 2 or 3 Dimensions
A E Hirst

Numbers, Sequences and Series
K E Hirst

Groups
C R Jordan and D A Jordan

Probability
J H McColl

Calculus and ODEs
D Pearson

In preparation

Analysis
E Kopp

Modular Mathematics Series

Statistics

A D Mayer and A M Sykes

*European Business Management School,
University of Wales, Swansea*

A member of the Hodder Headline Group
LONDON • SYDNEY • AUCKLAND

First published in Great Britain 1996 by
Arnold, a member of the Hodder Headline Group,
338 Euston Road, London NW1 3BH

© 1996 A D Mayer and A M Sykes

All rights reserved. No part of this publication may be reproduced or transmitted in any form or by any means, electronically or mechanically, including photocopying, recording or any information storage or retrieval system, without either prior permission in writing from the publisher or a licence permitting restricted copying. In the United Kingdom such licences are issued by the Copyright Licensing Agency: 90 Tottenham Court Road, London W1P 9HE.

British Library Cataloguing in Publication Data
A catalogue record for this book is available from the British Library

ISBN 0 340 63194 5

Typeset in 10/12 Times by
Paston Press Ltd, Loddon, Norfolk
Printed and bound in Great Britain by
J W Arrowsmith Ltd, Bristol

Contents

	Series Preface	vii
	Preface	ix
1	**Statistics – An Overview**	**1**
	1.1 Collecting data	1
	1.2 Data analysis	4
	1.3 Scales of measurement	5
	1.4 Applications of statistics	6
	1.5 Abuses of statistics	7
	Further exercises	9
2	**Exploring and Presenting Data**	**10**
	2.1 Frequency distributions	10
	2.2 Histograms	11
	2.3 Pie charts	13
	2.4 Stem-and-leaf displays	14
	2.5 Multiple displays	17
	2.6 Averages	19
	2.7 Statistics that measure dispersion or spread	25
	2.8 Boxplots	27
	2.9 Paired values	28
	2.10 Correlation	32
	2.11 Regression analysis	35
	Further exercises	38
3	**Probabilty Distributions**	**40**
	3.1 Probability, random variables and expected values	40
	3.2 The binomial distribution	42
	3.3 The Poisson distribution	48
	3.4 The hypergeometric distribution	54
	3.5 The geometric distribution	55
	3.6 The negative binomial distribution	56
	3.7 Continuous distributions	58
	3.8 The mean and variance of a continuous distribution	60
	3.9 The exponential distribution	61
	3.10 The normal distribution	63
	3.11 Sampling distributions	78
	Further exercises	79
4	**Estimation**	**82**
	4.1 A simple estimation problem	83
	4.2 Maximum likelihood estimation	85

Contents

- 4.3 Maximum likelihood estimation using a frequency table 88
- 4.4 Maximum likelihood estimation for common probability distributions 92
- 4.5 Maximum likelihood for normally distributed data 94
- 4.6 The method of least squares 97
- 4.7 Linear regression 100
- Further exercises 106

5 Hypothesis Testing **108**
- 5.1 The confidence interval approach 108
- 5.2 Classical formulation of hypothesis tests 109
- 5.3 Hypothesis tests for discrete data 113
- 5.4 Calculating the significance probability 114
- 5.5 The normal distribution with unknown variance 115
- 5.6 Matched pairs 118
- 5.7 Differences of means (two independent samples) 120
- 5.8 Non-parametric hypothesis tests 125
- 5.9 Tests for proportions 130
- 5.10 The distribution of the sample variance 132
- 5.11 Hypothesis tests in regression and correlation 135
- 5.12 The chi-squared goodness-of-fit test 142
- Further exercises 152

6 More Advanced Statistics **155**

Appendix A Datasets **157**

Appendix B Sampling from a Finite Population **162**

Solutions **164**

Index **173**

Series Preface

This series is designed particularly, but not exclusively, for students reading degree programmes based on semester-long modules. Each text will cover the essential core of an area of mathematics and lay the foundation for further study in that area. Some texts may include more material than can be comfortably covered in a single module, the intention there being that the topics to be studied can be selected to meet the needs of the student. Historical contexts, real life situations, and linkages with other areas of mathematics and more advanced topics are included. Traditional worked examples and exercises are augmented by more open-ended exercises and tutorial problems suitable for group work or self-study. Where appropriate, the use of computer packages is encouraged. The first level texts assume only the A-level core curriculum.

Professor Chris D. Collinson
Dr Johnston Anderson
Mr Peter Holmes

Preface

There are many different books on elementary statistics. Whilst most of them would substantially agree with each other on content, nevertheless they may appear distinctly different because of the variety of different approaches adopted, which in turn depend to some extent on the needs of the intended readers and their mathematical backgrounds. This book aims to provide a textbook for a student majoring in mathematics or statistics to support a first-year 24 lecture course. The reader should have a reasonable grasp of A-level mathematics including differential and integral calculus and be prepared to manipulate algebraic expressions involving summations, inequalities and the like. Indeed if any student of mathematics required justification for differential and integral calculus (the bread and butter of mathematics?), then statistics provides it *par excellence*.

With this assumed background, the book does not shy away from attempting to present a glimpse of the underlying mathematical structure of mathematical statistics, in order that the plethora of different statistical recipes in hypothesis testing (for example) should be less bewildering simply because they can be seen as the result of applying the same principle in different circumstances. These different circumstances stem from the choice of probability model thought to be appropriate to the physical context of the data to be analysed. The basic statistical paradigm starts with this physical context, models the context appropriately, applies standard statistical procedures in order to (a) validate the model chosen, and (b) make the required inferences.

Statistics is regarded by mathematicians as a branch of mathematics. Indeed it is. But because its aim is to uncover patterns amidst the chaos of randomness (a double chaos because there are so many different forms of randomness relating to different physical situations, as well as the inherent unrepeatability of randomness within any one of these situations), it is a very non-trivial branch of mathematics. You have to be prepared not only to accept the mathematical definitions of mathematical statistics, e.g. probability, probability density function, expected value, but also to recognise their sample equivalents (sample proportion, histogram, sample mean) and when you can swap easily between the two, you are in with a fighting chance of becoming a competent statistician, because you will not be so easily tempted to let the mathematical structure of your analysis run away with you beyond the bounds of its applicability in the context within which you are working!

Under two hundred pages is not a lot of space to cover the syllabus – there are many textbooks which justify twice this amount. So Chapter 1, setting the scene, is necessarily terse, but the chosen content is quite important (particularly the different types of measurements). Chapter 2 introduces many of the data analytic tools that are to be used subsequently, and Chapter 3 provides the basic probability models without getting side-tracked into the many fascinating probabilistic properties that you would expect if the book's title were *Probability and Statistics*.

In Chapter 4 we approach the topic of estimation and confidence intervals through the application of the method of maximum likelihood; later the method of least squares takes over because, for normally distributed data, they are equivalent. Consequently, estimation in one-sample, two-sample and regression problems is just a manifestation of applying the same principle. With confidence intervals firmly understood, the transition from estimation to hypothesis testing is relatively easy, and the final chapter is devoted to this.

Each chapter includes some illustrative examples and many exercises and suggestions for case studies. Some of these have been based on practical data-gathering exercises used in the first-year course in statistics at the University of Wales Swansea. They have proved to be valuable in making sure that the statistical analysis really does take place within a strong physical context – each student does his or her own experiment, introducing his or her own influence on the results, so that there is no unique answer expected!

Finally we must thank those who have helped this book along its way: the staff of Arnold and Peter Holmes, the staff of the European Business Management School of the University of Wales Swansea for academic and technical computing support, Adrian Smith of Causeway for his excellent APL Postscript Graphics facilities, our parents for their unflinching support when we were students, and above all, our wives Kathie and Claire for their patience and forbearance throughout it all.

Alan D. Mayer
Alan M. Sykes
July 1995

1 • Statistics – An Overview

To most people the word *statistics* conjures up an image of pages upon pages of numbers. Some will think of batting averages or league tables; others, sales figures or share prices; yet others, examination marks, 'SATs' results or IQ scores. In each case the 'statistics' generally refers to the results of collecting and compiling and possibly summarising the raw data collected. Collecting, tabulating and summarising data is an important part of the science of statistics. Equally important, and more of an art than a science, is the task of discovering and interpreting patterns and structure in the data. The purpose of it all is to try to find answers to difficult questions.

1.1 Collecting data

At first sight, collecting data would seem to be the easiest of a statistician's tasks. At its simplest it may involve a visit to the local library. At the other end of the scale it may necessitate a 10 year expedition to Antarctica! We rarely produce statistics for their own sake – we collect them because someone somewhere sometimes needs information contained in them for a particular purpose.

What information do we really need?

Suppose we want to make a statement about the ages of students at our college. One possibility would be 'the average age is 20.35 years'. To obtain that figure (or to estimate it reasonably accurately) we need to know the ages of all of the students. If we have access to a computerised database with all the students' dates of birth, we can probably extract the figure we want in minutes. If we have to send a questionnaire to the students the job could take weeks – and almost certainly not all would reply.

We may be satisfied by an alternative form of statement such as 'most of the students are between 18 and 22 years of age'. We can obtain this information by a show of hands ('How many of you are over 22 years of age?') in four or five lecture theatres.

Don't re-invent the wheel

Collecting data is time consuming. It is often expensive and sometimes destructive (where the unit being measured is damaged or destroyed by the test). People who are asked the same question too many times get angry. All of these are good reasons to check whether the study you are about to embark on has already been done by someone else.

Data collected from our own original sources for the purposes of the study in hand is called **primary data**. It has the advantage that we are usually able to control precisely what is collected and can monitor, if not influence or control, extraneous conditions which may affect the data.

Data collected and compiled for another purpose, but which we can utilise for our own study, is called **secondary data**. We must treat it with caution, as we may not know the precise conditions under which it was collected, but it is less time consuming and often cheap or even free. In some cases, such as national census returns, there is far more data available than we could ever hope to collect ourselves.

The easiest data to handle is obtained from computerised databases, on-line or CD-ROM. More and more sources are becoming available on the Internet: for example, StatLib, a library of statistical software, datasets and information, held in the Higher Education National Software Archives (HENSA), based at the University of Kent at Canterbury (http://www.hensa.ac.uk/ftp/mirrors/statlib/). These sources have the advantage that large datasets can be downloaded and fed to packages such as Minitab, SPSS, SAS and so on.

Other more traditional sources are published papers, books, periodicals, university dissertations, theses and working papers. If the data on which the papers are based are not published, the authors are often willing to release it for research purposes. Don't forget to talk to colleagues who may be doing similar work in your own organisation!

Most libraries now have computerised search facilities, subject and author indexes, citation indexes and so on. It is well worth while finding out how they work.

Observational and experimental data

In a study to determine how fit they are, various people are asked to run around a field and then their heart rate, blood pressure, etc., are recorded. Other variables such as age, height, weight are also recorded. Suppose we discover that fat people are less fit than thin people. Does this mean that being fat makes you less fit? If they ate less and lost weight would they be fitter? Or is the problem that they do not do enough exercise, and as a result (a) they get fat, and (b) they are not fit?

The difficulty we face in this study is that we cannot control who is fat and who is thin. We can only *observe* the association between fatness and fitness. The data collected in such a study is called **observational data**.

If the people volunteered to be allocated at random (or perhaps by some design) to two groups, one of which would be starved and the other overfed to produce a thin group and a fat group, we would be able to investigate whether being fat because of overeating *caused* unfitness. Data from a study in which we have control over which subjects receive a treatment, and which do not, is called **experimental data**. Of course, it may be difficult to persuade people to volunteer for such an experiment!

We can discover *association* between variables from observational data; experimental data may allow us to dig deeper and determine *cause and effect*.

Sampling

In a statistical study we must ensure that we identify clearly who or what are the objects of our study. The group so defined is called the **population**.

If we want to investigate the difference between men's income and women's income, by household, in a small village, we might consider asking every inhabitant for the relevant data. When we collect data on every member of the population, the result is a *census*. A similar exercise applied to the whole country is called a National Census and is carried out by the Government from time to time. Unless the population is small, a census is usually prohibitively expensive.

The alternative to a census is to survey a **representative sample**, which is a subset of the population, selected for analysis. The method of selection determines how well the population is represented by the sample.

One of the most common sampling techniques is the **random sample** in which every member of the population has an equal chance of being in the sample. The selection of winning numbers for the National Lottery is a random sample.

To estimate the average total length of 18 hole golf courses in Britain, using a random sample of size 20, you would have to list all the 18 hole golf courses (the population), select a random sample of 20 of them (e.g. using random numbers from tables or from a computer) and telephone each of those selected to obtain the information. The average of the sample of 20 would then be an *estimate* of the population average.

A more sophisticated form of random sampling is called **stratified random sampling**. Suppose the committee of the local tennis club wants to sound out its members on plans for improvements to the club. Its membership of 150 is made up of 100 men and 50 women, and they plan to talk to a sample of 30 people. A random sample may produce a majority of women, or no women at all. To avoid the possibility of such bias, the committee takes a stratified random sample by selecting 10 women at random from the 50 women members and 20 men from the 100 men. In other words, the *proportions* of men and women are the same in the sample as in the population. This sort of stratification can be performed on several variables simultaneously.

If items to be sampled are produced in boxes of 10, it is often convenient to select a number of boxes and test every item in each selected box. This is called **cluster sampling**. It is acceptable if items are allocated more or less randomly to boxes at the packing stage. If, however, they are packed in the order they come off the production line, it is likely that any items, which are defective due to a production line malfunction, will be in one or two boxes only. If those boxes are not selected, the defective items will be missed altogether.

One way of getting a sample of 10% of a population is to knock on every 10th door, or select every 10th member from the membership list and so on. This is called **systematic sampling**.

Methods which should be avoided, or treated with extreme caution, are **accessibility sampling** where items are chosen for ease of access (e.g. the bricks near the top of the pile, or employees who live within two miles of the office), and **judgemental sampling** where items are deliberately chosen to balance the sample in the view of the person doing the sampling. Both of these methods tend to introduce bias in one form or another.

EXERCISES ON 1.1

1. Explain why the following samples are not representative:
   (a) A sample of 50 school teachers in Birmingham is taken to estimate the average salary of people who live in Birmingham.

4 Statistics

    (b) In order to estimate the average number of books read each year by teenagers in Cardiff, the City Librarian interviews a random sample of teenagers visiting the Central Library.
    (c) To estimate the number of dyslexic pupils in a school, all those whose surnames begin with the letters A–E are tested.
    (d) A political canvasser questions everyone he or she can find at home between 2 and 4 p.m. on a Wednesday afternoon.
2. In a committee of five persons, Andy, Bernie and Carol intend to vote for the motion, Donna and Esmail intend to vote against. I select two committee members at random and ask which way they will vote. How many different selections of two could I make? In how many of these would the result be one vote for and one vote against?

Sample surveys

These are a major source of data for market research purposes. They can be conducted in a variety of forms. **Personal interviews** are the most effective, allowing the interviewer to be flexible in meeting unexpected responses, but they are very expensive and time consuming, and it is easy for the interviewer to bias the results. **Telephone interviews** are much cheaper, but many numbers are unlisted (ex-directory) which introduces a bias to the sample. **Mail surveys** are also low cost, but the response rate tends to be very low, unless suitable incentives can be offered to encourage response. Many researchers opt for a combination of several methods.

1.2 Data analysis

After collecting data we have a variety of tools available to analyse it. We use **summary statistics** such as averages to summarise the raw data, and **charts** and **graphs** to display it. Some of these techniques are described in Chapter 2.

    To analyse and interpret the data we use **inferential statistics**, and these are described in Chapters 3–5.

Exploratory data analysis

When a set of data has been collected for a specific purpose, e.g. to test whether a new drug is effective, we are often tempted to perform the agreed test using a computer package, and never actually look at the data. This strategy means that we throw away an opportunity to gain insights into the structure of the data and risk missing information relevant to the test.

    The idea of **exploratory data analysis** (EDA), a set of techniques pioneered by John Tukey in the 1960s, is that the data should help us to select the analysis procedures, or modify the procedures previously planned. EDA requires a flexible approach, an 'open mind' and a willingness to depart from rigid, pre-determined procedures.

Deduction and induction

In Chapter 3 we discuss various probability models for data. We can make statements like 'if our model is correct, then these results will follow'. This is deductive reasoning, arguing from the general to the specific (from the population to the sample). Statistical inference, described in Chapter 5, involves the reverse argument: 'from our observed sample, we conclude the following about the population'. This is **inductive reasoning**, from the specific to the general (from the sample to the population).

1.3 Scales of measurement

We use numbers to mean a variety of different things. As a result, it is easy for us to confuse the meanings and use inappropriate methods of analysis. For example, consider the numbers 1 and 2. We may say that 2 is twice as big as 1; the difference between them is 1; their average (mean) is 1.5 and so on. Depending upon the reason for allocating the numbers 1 and 2, some or all of these statements may be false. We discuss here four scales of measurement, all of which are commonly used. They are presented in order, 'weakest' first.

The nominal (classificatory) scale

In a database, we may use 1 and 2 to represent male and female. Clearly female is not twice male, or even 'greater than' male (at least, not by virtue of the numbers allocated!); the difference between male and female is not 'male'; the average is not 1.5. On the **nominal scale**, numbers are used only as convenient *labels* to classify the data. Not even their order has any meaning.

The ordinal (ranking) scale

We often ask people to express their opinion of something by giving it a number (e.g. from 1 to 5, where 1 may be 'very bad' and 5 'very good'). Alternatively, we may perform a test to destruction on a number of items, and record only the *order* in which they fail. On the **ordinal scale**, the order is meaningful (2 is 'better' than 1; or 2 lasts longer than 1) but the difference between 1 and 2 is not necessarily the same as the difference between 2 and 3. The median (see Chapter 2) can be used for an ordinal scale, but the mean is inappropriate. On this scale, if $x > y$ and $y > z$, it follows that $x > z$.

The interval scale

On the **interval scale** the *differences* between values are meaningful, but the *ratios* are not; there is a *unit of measurement* but the location of zero on the scale is arbitrary. For example, on the Fahrenheit or Celsius temperature scales, zero does not mean that there is no heat present. 20°C is 10 degrees warmer than 10°C, but it is not twice as warm. Another example is time: 10 a.m. is not twice as late as 5 a.m. Both median and mean are appropriate on this scale.

The ratio scale

An interval scale, with the addition of a true zero point, becomes a **ratio scale**. For example, 5 kg of potatoes weigh 4 kg more than 1 kg of potatoes, and they also weigh *five times* as much. This is the 'strongest' scale of measurement.

6 Statistics

Requirements for hypothesis tests

The tests discussed in Chapter 5 require certain minimum levels of measurement. The parametric tests, such as t-tests, require data that is at least interval. The non-parametric tests, such as Wilcoxon signed rank and Mann–Whitney, require ordinal data.

Summary

Scale	Characteristics			Operations
	Order	Distance	Origin	
Nominal	✗	✗	✗	Equality
Ordinal	✓	✗	✗	Greater than or less than
Interval	✓	✓	✗	Equality of differences
Ratio	✓	✓	✓	Equality of ratios

EXERCISES ON 1.3

1. What scale of measurement is most suitable for the following?
   (a) Ranks of personnel in the armed forces.
   (b) Telephone numbers.
   (c) Shades of the colour green in a paint chart.
   (d) Marks in a statistics examination.
   (e) Sizes of paper (A1, A2, A3, etc.).
   (f) Scales of measurement for data.
2. Five members of my athletics club took part in a race and came 1st, 4th, 7th, 8th and 10th. Which of the following are appropriate to quote?
   (a) The highest placing was 10th.
   (b) The mean position was 6th.
   (c) The median position was 7th.

1.4 Applications of statistics

'No argument is complete without the figures to back it up.' This rather dubious dictum rules much of the business world, and a good deal of modern science. The truth behind the dictum is that modern statistics is one of the greatest discoveries of the twentieth century, and is used for good or evil in almost every area of human endeavour. Used correctly, statistics can result in changes which are of great benefit. Manipulated by the unscrupulous, statistics can be used to 'justify' the most outrageous claims. A small selection of (potentially) beneficial applications includes:

- Medicine: testing of new drugs; prediction and control of epidemics; detection of causal links such as smoking and cancer.
- Forensic science: e.g. using distributions of properties such as the refractive index of glass as evidence that a suspect has visited the scene of a crime.

- Legislation: data on seat-belts, drinking habits, etc., are used to justify changes in the law.
- Education: provision of facilities, based on projected demand (including closing or opening of schools on the basis of population projections).
- Politics: opinion polls are conducted daily on any issue which will help to sell a newspaper or increase the ratings of a radio or television programme.
- Finance: changes in indexes such as the FT-SE 100, Dow Jones, etc., *reflect* trading on the Stock Exchanges, but also *cause* panic buying and selling.
- Meteorology: one of the earliest and most highly developed sciences of forecasting (what would we do without the weather person?).
- Agriculture: the sophisticated control over the production of food exercised by all developed countries is driven by statistics – without statistics the European Union would cease to exist!
- Geology: location of oil, gas and coal fields, assessment of their size and resources.
- Sociology: studies of people's attitudes using sample surveys.
- Industry: market research, again using sample surveys, and quality control.

The last topic, quality control, has revolutionised attitudes in the business world. In essence it is the statistical monitoring of production processes with a view to improving consistency and quality in the end product. Pioneered in the United States by Walter Shewhart and Edwards Deming, the methods were adopted by Japanese industry soon after the Second World War, and were largely responsible for Japan's miraculous recovery from the ruin of the war. Rather belatedly, the United States, Europe and others are recognising the importance of statistics in this field.

1.5 Abuses of statistics

'There are three kinds of lies: lies, damned lies, and statistics' (Benjamin Disraeli). Distrust of statistical results, and abuse of statistics, go back a long way into history. The abuse can occur at three levels:

- Fabrication of data or reported statistics.
- Dishonest statistical practices, e.g. selective reporting.
- Strictly accurate displays or assertions couched in terms intended to mislead.

To these we may add genuine errors, such as typing or data-entry errors, invalid assumptions leading to inappropriate statistical tests, and so on. We shall restrict our attention to the malicious variety.

Misleading displays or assertions

One of the most common ways in which we can use a graph or chart to mislead the reader is by omitting any reference to the origin. Figure 1.1 shows the average salaries for employees of two firms over the last four years. Which firm would you join?

Of course, the two graphs display exactly the same figures. With all displays, **look at the numerical information: do not rely on general impressions**.

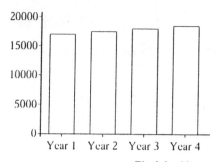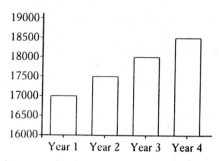

Fig 1.1 Histograms of average salaries.

Sometimes we use different size pictures to represent different quantities, e.g. bank notes to represent company profit. If company A makes twice as much profit as company B, it should be represented by a bank note which is twice as big. If we double its length *and* width, the resulting area is *four* times the size and greatly exaggerates the apparent difference between the companies.

'Eight out of 10 shoppers who answered our questionnaire preferred our brand.' This sounds good – but what if only 10 people answered the questionnaire? Even worse, what if eight of them were our own employees?

'Of the washing machines sold by our company in the last 20 years 90% are still in daily use.' If you bought a washing machine from this firm, you might think there is a good chance that it will last for 20 years. What the firm 'forgot' to mention is that sales have increased considerably over the years, and only 5% of the machines were sold more than 10 years ago, and none of them are still in use. In fact the *average* life is about 5 years.

Selective reporting and data dredging

The researcher who conducts 100 experiments but reports only the five whose results are favourable to his or her pet theory is clearly being dishonest. The political canvasser who decides to restrict a survey to areas where voters are traditionally favourable to his or her party is being equally dishonest, but may claim that, as he or she did not *collect* the unfavourable data, he or she cannot be accused of discarding it. Also, in this area of work, the precise wording of the question can be of vital importance – a small change can lead to totally different results.

What about the researcher who runs a large dataset through every test in a computer package, making a note of 'anything significant', and then reports 'the interesting ones'? This is another form of selective reporting. There is nothing wrong with 'exploring' a set of data, but the exploration should be directed and intelligent. Indiscriminate 'data dredging' leads to a large number of spurious and potentially misleading results. A set of data is in some respects like a human being: if you subject it to a prolonged interrogation and flog it half to death it may well tell you anything you want to hear.

EXERCISES ON 1.5

1. Criticise the histogram in Fig 1.2.

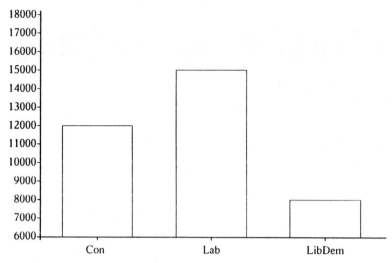

Fig 1.2 Votes cast for three major parties.

FURTHER EXERCISES

1. Discuss the variables in the cholesterol dataset (Dataset 1, Appendix A) and classify them. In particular, determine which variables are ratio-scale variables.
2. Find out the current world population figures, say for the 10 largest countries. Present this information (a) as a bar chart, (b) relative to the population of Britain, (c) as a pie chart (see Chapter 2).
3. Good quality newspapers often include pictorial representations of data. Collect 20 such examples and discuss their good and bad points.

TUTORIAL PROBLEM

Your local authority wishes to estimate how many lamp-posts it has. Consider different methods you might employ which require a minimum of resources (time and money) to do this. A good method should enable you to derive not only an estimate, but some idea of how accurate that estimate is likely to be.

2 • Exploring and Presenting Data

After collecting the data, the next step is to summarise the main properties and look for patterns and relationships. Numerous methods have been devised to help with this task, and one of the most difficult decisions for the inexperienced is the choice of the most appropriate techniques for the particular problem in hand. The methods fall broadly into two categories: numerical summaries called statistics, and pictorial representations such as graphs and charts; of course, these interact!

Before choosing a particular approach, we must consider the purpose of the analysis. Sometimes the aim is to *explore* the data – look for any patterns or relationships, and allow the data to determine the form of the analysis. This exercise is called **exploratory data analysis** (EDA). Very often, however, the study is motivated by interest in specific relationships which are suspected to exist, so that the methods of analysis may be dictated by the objectives. This approach is called **confirmatory analysis**. In either case, it should be borne in mind that the final stage of the project will be to present the results to other people, verbally, or with illustrations, or in writing.

The content of the final presentation will depend as much upon the nature of the audience as upon the points to be illustrated. For people familiar with the form of the data, numerical summaries can be quite adequate, but charts and graphs are generally more effective. There is room at this stage for more inventive presentations. For example, we may represent annual UK oil production by a chart made up of pictures of barrels of oil. We must take care, however, to ensure that it is clear to the reader whether it is the heights of the barrels, the areas of the barrels on the page, or the implied volumes of the barrels that are proportional to annual UK oil production.

2.1 Frequency distributions

Dataset 1 (Appendix A) contains data from a study of plasma cholesterol levels in 118 patients before and after a dietary and lifestyle intervention programme. The cholesterol level after intervention (*chola*) is ratio data, recorded to two decimal places, and a glance down the column will be sufficient to see that most of the values are between 5 and 8. Beyond that, it is difficult to digest the information in its raw form. A useful technique is to group the values into classes defined by intervals, and count the number of values in each class. The result is a **frequency distribution**, and an example is shown in Fig 2.1.

This is a considerable improvement on just presenting the raw data. In this case, the original 118 data points have been reduced to 11 category descriptions and 11 frequencies. The frequency distribution shows that the majority of values are

Category	Frequency
3.51–4.00	1
4.01–4.50	0
4.51–5.00	2
5.01–5.50	13
5.51–6.00	7
6.01–6.50	22
6.51–7.00	24
7.01–7.50	30
7.51–8.00	10
8.01–8.50	6
8.51–9.00	3

Fig 2.1 Frequency distribution of cholesterol after intervention.

between 6 and 7.5, the minimum value is in the interval 3.51–4 (and is somewhat isolated from its nearest neighbour) and the maximum is in the interval 8.51–9.

The classes are described in terms of the **class limits** – the smallest and largest values that can be allocated to the class. For example, the first class has class limits 3.51 and 4.00. The values are recorded to two decimal places, and the limits must be chosen to ensure that every observable value can be allocated to one and only one class. Class limits 3.5–4.0, 4.0–4.5, 4.5–5.0, ... would therefore be unacceptable, as the value 4.00 could be allocated to either of the first two classes. On the other hand, class limits 3.50–3.99, 4.00–4.49, 4.50–4.99, ... would be quite acceptable, and may lead to a slightly different frequency distribution.

For continuous data, it is important to note that a value such as 4.01 actually represents, after rounding, any value from 4.005 up to (but not including) 4.015. So a class labelled 4.01–4.50 really includes values from 4.005 to 4.505. These are called the **class boundaries**.

The **class interval** is the difference between adjacent class boundaries (here 0.5), and the **class representative** is the mid-point of the class (half way between the class boundaries). The class representatives for Fig 2.1 are 3.755, 4.255, 4.755, ...

2.2 Histograms

The information presented numerically in a frequency distribution may be displayed pictorially in a **histogram**. The frequencies are represented by rectangles, either vertical or horizontal, the area of each rectangle being proportional to the corresponding frequency. The histogram of the frequency distribution in Fig 2.1 is shown in Fig 2.2.

Note the scale on the vertical axis. Some statisticians insist that these frequencies should be divided by the width of the category (in this case 0.5). This is essential if there are categories of different widths in the same histogram, but the more common computer packages use the raw frequencies (as in Fig 2.2).

When the histogram is drawn horizontally, it is usual to label each category with its middle value. Strictly this should be the class representative, but in practice it is usually rounded.

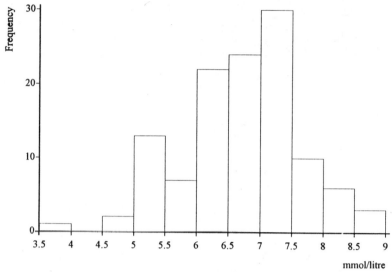

Fig 2.2 Histogram of *chola*.

Choice of classes

What factors should we consider when choosing the class limits? The first thing to realise is that there are no hard and fast rules. It is usual, whenever possible, to choose equal class intervals, and the number of classes should be large enough to reveal details in the structure of the data, but small enough to ensure that at least some classes contain a reasonable percentage (perhaps 10% or more) of the data points. Typically there are between five and 15 classes. Finally, the class limits should be based around whole numbers or 'sensible fractions' such as halves or quarters.

Computer packages

To produce a frequency distribution or histogram on a computer, the data must be provided, and the classes must be specified. Most packages will choose classes for you, but the choice may or may not satisfy your requirements. If the choice is unsuitable, you must specify the classes yourself, by means of parameters. Provided all class widths are equal, the classes can be fully described by two parameters: (i) the class width and (ii) either a class representative or a class boundary.

Example 1

Calculate suitable classes for a frequency distribution of *wtb* in Dataset 1 – the weights of the subjects before intervention.

First decide on an approximate number of classes, k, say 10. Then extract from the data the maximum and minimum values. For *wtb*, these are max = 124 and min = 47. Next, calculate

$$\frac{\max - \min}{(k-1)} = \frac{124 - 47}{9} = 8.556$$

This is the required class width to produce 10 classes with the minimum and maximum values at the centre of their classes (why?).

As the result is neither an integer nor a convenient fraction, it should be adjusted up or down. A class width of 8 will give more than 10 classes, while 9 will give fewer. Another good candidate is 10, starting at 45.

EXERCISE ON 2.2

1. Calculate the frequency distributions, and draw the corresponding histograms, for the three class structures suggested in Example 1. If you have access to a computer package, learn how to produce histograms, and experiment with different class structures for *age*, *ht*, *wtb*, *wta* and *chola* in Dataset 1.

Large datasets

Sometimes, particularly when the number of observations is very large, datasets are recorded only as frequency tables. Dataset 3 is such a table, showing the population of Ireland in 1841, by sex and age. Figure 2.3 is a histogram of this data, combining the two sex categories.

2.3 Pie charts

Figure 2.3 shows that the numbers in each age category decline fairly consistently as age increases. Fine details can also be seen – for example, there are fewer than

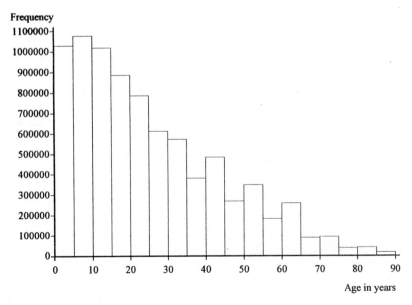

Fig 2.3 Histogram of the population of Ireland in 1841.

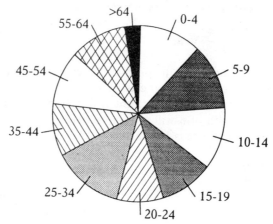

Fig 2.4 Pie chart of the population of Ireland in 1841, by age.

may be expected in the 35–39 age group. It is not easy, however, to see from this diagram that the 0–9 year olds make up about 25% of the population. For this, another pictorial representation, called a pie chart, is more suitable.

A pie chart is a circular depiction of data where the whole 'pie', representing 100% of the data, is divided into 'slices' whose sizes are proportional to the sizes of the categories they represent. Figure 2.4 is a pie chart for Dataset 3, the population of Ireland in 1841. Note that the five small categories representing the over-65s have been combined for convenience of representation.

That about 25% of the population are 0–9 year olds is now immediately apparent, as are the facts that about 50% are under 20 years of age, about 25% are in the 5–14 range, and so on.

2.4 Stem-and-leaf displays

The stem-and-leaf display is a type of horizontal histogram which is particularly appropriate for EDA. At its best, it preserves full details of the values of the raw data, in a histogram made up of numbers, but some datasets are better suited to this treatment than others.

The technique is best illustrated by an example. Consider the 'weight before intervention' (*wtb*) column of Dataset 1 – the data values range from 47 kg to 124 kg. These can be conveniently categorised, for the purposes of a frequency distribution, into classes 40–49, 50–59, 60–69, ... , i.e. 40s, 50s, 60s, ... The values 4, 5, 6, ... (the 'tens') become the stems for the stem-and-leaf display, and the 'units' become the leaves. This leads to the display illustrated in Fig 2.5.

Note that the leaves on each stem have been recorded in ascending order, so the display can be read as an ordered listing of the dataset. Each leaf digit represents an individual value, so the overall shape of the stem-and-leaf display is equivalent to a histogram of the same data. The first column of numbers is a cumulative count of the number of values in that row or on rows towards the nearer edge of the distribution. The brackets identify the row containing the middle value (or middle two values) and the value in brackets is the number of values in that row.

```
    1     4 7
   17     5 2334455566677999
   40     6 00222233566666777888899
  (34)    7 0011112222333334444555566666778899
   44     8 000001222244455566777789
   20     9 111122344578889
    5    10 027
    2    11 7
    1    12 4
```

Fig 2.5 Stem-and-leaf display of *wtb*.

To produce a sensible number of categories, and to preserve the property that each leaf is one digit, some data may have to be rounded before being categorised in this way. For example, a stem-and-leaf display of *chola* from Dataset 1 is shown in Fig 2.6.

```
    1     3 9
    3     4 57
   23     5 01233334444445667779
  (46)    6 0011112222223333334444555566777788888888899999
   49     7 000000000111122223333344444445666778889
    9     8 001123679
```

Fig 2.6 Stem-and-leaf display of *chola*.

Here the data values have first been rounded to one decimal place, and then the first decimal place has been taken as the leaf, with the 'units' as the stems. The result, with only six categories, is less than satisfactory. An improved version of the stem-and-leaf display for this data is achieved by a process called splitting the stems. This involves producing more than one row for each stem, e.g. a row for leaves 0–4 and a row for leaves 5–9. The result of splitting the stems for this data is shown in the form of Minitab output in Fig 2.7. With 11 categories, this is a more reasonable representation of *chola* and corresponds quite well to Fig 2.2.

```
MTB > stem-and-leaf 'chola'
  Stem-and-leaf of chola      N = 118
  Leaf Unit = 0.10

      1      3 9
      1      4
      3      4 57
     16      5 0123333444444
     23      5 5667779
     45      6 001111222223333334444
    (24)     6 555566777788888888899999
     49      7 000000000111122223333344444444
     19      7 5666778889
      9      8 001123
      3      8 679
```

Fig 2.7 Minitab output, splitting the stems for *chola*.

To produce an acceptable stem-and-leaf display of the *age* column in Dataset 1, we must take the splitting of the stems a stage further. A first effort at the stem-and-leaf display for the ages is shown in Fig 2.8.

```
   4    2 4679
  44    3 112334444555666777778888888888999999999
 (41)   4 000001222222333344444566666777788888899
  33    5 0011111222333334444445555555678
```

Fig 2.8 Stem-and-leaf display of *age*.

Clearly there are far too few categories – even doubling them will not be sufficient. Here we can split the stems to produce five rows per stem, with leaves 0–1, 2–3, 4–5, 6–7 and 8–9. The result, which is a far more informative display, is shown in Fig 2.9.

```
MTB > stem-and-leaf 'age'
Stem-and-leaf of Age      N = 118
 Leaf Unit = 1.0
      1     2 4
      3     2 67
      4     2 9
      6     3 11
      9     3 233
     16     3 4444555
     25     3 666777777
     44     3 8888888888999999999
     50     4 000001
    (10)    4 2222223333
     58     4 4444445
     51     4 666667777
     42     4 888888999
     33     5 0011111
     26     5 222333333
     17     5 44444445555555
      3     5 67
      1     5 8
```

Fig 2.9 Stem-and-leaf display of *age*, with five rows per stem.

EXERCISE ON 2.4

1. The values below are systolic blood-pressures of patients admitted to a hospital. Construct a stem-and-leaf display of this data.

112.1	138.6	115.9	109.5	108.2	110.9	159.6	115.8	122.3	122.4
136.5	123.8	117.5	130.4	137.3	105.5	132.8	150.2	154.3	107.3
133.7	121.4	105.9	122.2	97.1	177.7	98.1	125.7	118.0	128.2
109.8	118.7	123.8	117.2	121.1	146.2	154.7	168.3	125.9	110.2
123.8	156.1	130.1	121.3	112.3	144.2	127.8	144.9	130.6	127.8

117.3 117.2 118.2 186.3 113.7 147.7 138.0 124.5 143.4 122.6
133.2 117.6 119.9 121.1 149.1 138.2 153.9 120.5 100.6 162.9

2.5 Multiple displays

Two of the variables in Dataset 1 are categorical, and have only two possible values – the columns representing the sex of the patients and the type of consultant who treated them. These variables may be used to split the other columns of data into subsets. For example, it may be useful to contrast the men's 'after intervention' cholesterol levels with those of the women. Separate histograms may be produced for each subset, and most packages allow for this option. In Minitab, for example, the *by* parameter will create a separate histogram from each value in the *by* variable, so the command sequence shown in Fig 2.10 will produce a histogram of *chola* for male patients, followed by a histogram of *chola* for female patients.

```
MTB > histogram 'chola';
SUBC> by 'sex'.
```

Fig 2.10 Minitab commands to produce histograms of *chola* by gender.

The contrast between the subsets is illustrated more effectively by combining the two histograms. This can be achieved in several ways, two of which are illustrated below: Fig 2.11 is a 'back-to-back' histogram, which makes comparison of the distributions very easy, but can only be used for two categories; Fig 2.12 is a histogram of the combined data in which the contributions of each category are distinguished by different symbols, colours or shadings. The distributions look very similar, but there are more men with very high cholesterol levels than women, and more women with very low levels than men.

```
                    Male              4   *         Female
                                     4.5  *
                             ***      5   *
                      *******        5.5  ********
                      ********        6   *****
                       ******        6.5  **************
              ****************        7   ****************
                      *********      7.5  **********
                           *****      8   ****
                             ***     8.5  *
                               *      9
```

Fig 2.11 Back-to-back histogram of *chola* by gender.

As a second example consider Dataset 2, the times recorded by medal winners in the Olympic Games 4 × 100 metres relay, 1928–1988. We can divide these values into subsets in many ways, e.g. by gender, by medal (gold, silver, bronze), and by such categories as 'pre-Second World War' and 'post-Second World War'. Figure 2.13 is a histogram of all the recorded medal times.

18   Statistics

```
    4       F                         F  Female
    4.5     F                         M  Male
    5       FMMM
    5.5     FFFFFFFFMMMMMM
    6       FFFFFMMMMMMMM
    6.5     FFFFFFFFFFFFFMMMMMM
    7       FFFFFFFFFFFFFFFFMMMMMMMMMMMMMMMM
    7.5     FFFFFFFFFFMMMMMMMMM
    8       FFFFMMMMM
    8.5     FMMM
    9       M
```

Fig 2.12   Character histogram of *chola* by gender.

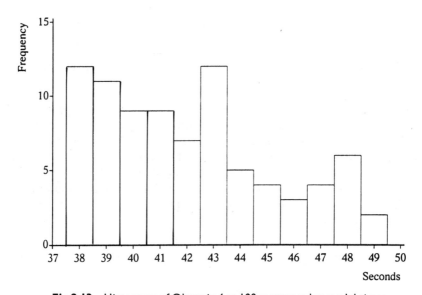

Fig 2.13   Histogram of Olympic 4 × 100 metres relay medal times.

An obvious, rather unusual property is the relative sparseness of values near the middle of the range (around 42 seconds). The reason for this is immediately apparent when we consider a histogram of the data, categorised by gender (Fig 2.14).

It is now clear that almost *all* the men's times are smaller than almost *all* the women's times, and also that the range of the men's times is smaller. This dataset is listed in an orderly manner, and the solution of the 'problem' is obvious. Often data of this sort arrives in a less logical order, and the division into categories is only apparent from graphical displays.

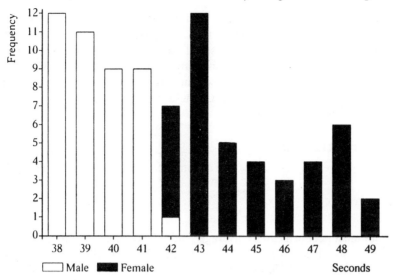

Fig 2.14 Histogram of Olympic 4 × 100 medal times by gender.

EXERCISES ON 2.5

1. The following data are the number of three-letter words found on each of 80 pages of text:

```
 62  119   92   98   70   55  112  112  135   85
 97  125   54   55   98  111   51   85   57   88
112  104  134  127   98   59  109   88  114  132
119   74   55  117   80  107  119  140   83   73
119   96   98   89   37   87  110   55   70   99
 73   52   55   63   45   74  111  112   36  112
 76   77   59  119   75   54   39  116   37   76
 65   55  113   78   72  115   35   99  100  105
```

   (a) Illustrate the data by means of a stem-and-leaf diagram.
   (b) You are now given further information: the first 40 pages are from an adult science fiction book, and the remainder are from a children's adventure book. Compare the two distributions by means of a back-to-back histogram. Is the result as you would expect it to be? Can you offer an explanation for what you have observed?
2. Using Dataset 1 (Appendix A), produce a back-to-back histogram of the weights after treatment (*wta*) by gender.

2.6 Averages

The visual representations used so far in this chapter are ideal aids to word descriptions of the data, but for many purposes brief numerical summaries (called **sample statistics**) are more appropriate. For example, when comparing the men's and women's cholesterol levels in Fig 2.12, it would be useful to represent each set of data by a small number of values, so that a simple comparison can be made.

The briefest possible summary of a set of data is a single value which may be considered in some way to be representative of all the values. The precise requirement may be expressed in phrases such as 'middle value', 'most common value', 'centre of gravity', and so on. These representative values are called averages or measures of location.

The arithmetic mean

For a set of n values, $x_1, x_2, \ldots, x_n$, the arithmetic mean, denoted $\bar{x}$, is defined by

$$\bar{x} = \frac{1}{n}(x_1 + x_2 + \cdots + x_n) = \frac{1}{n}\sum_{i=1}^{n} x_i$$

This is what most people mean by 'average' – add the numbers up and divide by n – but is it a reasonable value to use as a summary for a whole dataset? As an example, consider the cholesterol data. First calculate the mean of the *chola* column (see Fig 2.15 for Minitab output).

```
MTB > mean 'chola' k1
    MEAN     =      6.7281
MTB > let c9=('chola' lt k1)
MTB > sum c9
    SUM      =     53.000
```

Fig 2.15 Minitab session on *chola*.

The result is 6.7281. But this is not the 'middle' of the data, as the output shows that 53 values are less than the mean. Clearly the mean is not the middle in the sense that half the values are below it and half the values are above it.

Repeat the procedure with *cholb*, cholesterol before intervention (see Fig 2.16).

```
MTB > mean 'cholb' k2
    MEAN     =      7.3525
MTB > let c10=('cholb' lt k2)
MTB > sum c10
    SUM      =     66.000
```

Fig 2.16 Minitab session on *cholb*.

Now 66 values are below the mean (and hence 52 values are above) – an even more extreme result, and in the other direction! Look at the histograms of the two sets of data, side by side, in Fig 2.17.

The *chola* histogram may be described as 'a gradual rise followed by a rapid fall', whereas *cholb* is 'a rapid rise followed by a gradual fall'. This effect is known as **skewness** – *chola* is skewed to the left (longer left tail) and *cholb* is skewed to the right (longer right tail).

Exploring and Presenting Data   21

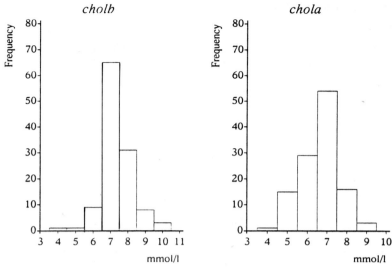

Fig 2.17   Histograms of cholb and chola.

To appreciate the true interpretation of the mean, imagine cutting the shape of a histogram from a piece of card, and attempting to balance it on a knife-edge, with the knife-edge parallel to the vertical axis – the point at which the histogram balances is the mean. In other words, representing the values by equal masses, the moments of the masses about the mean must be in equilibrium, so the distances from the mean must sum to zero.

To confirm this, subtract the mean from all the values and sum (see Fig 2.18).

```
MTB > let c11='chola'-k1
MTB > sum c11
    SUM       =  0.000013113
```

Fig 2.18   Subtracting the mean and summing.

The fact that this is not quite zero serves to remind us that there are round-off errors involved in this type of computation. Algebraically, it can be seen that SUM should indeed be zero:

$$\sum_{i=1}^{n}(x_i - \bar{x}) = \sum_{i=1}^{n} x_i - n\bar{x} = \sum_{i=1}^{n} x_i - \sum_{i=1}^{n} x_i = 0$$

As a 'centre of gravity' the mean is an effective representative of sets of data which cluster reasonably symmetrically about a central value, but can be rather misleading in some circumstances. The following four datasets all have 10 values with mean 5.0. Their histograms are illustrated in Fig 2.19. The first is closely clustered around 5, and the second, though more widely spread, is still quite homogeneous. In contrast, the third dataset is really made up of two distinct sets of points, and the mean lies in the central region between the two sets, and the fourth dataset possesses a single, very large 'outlier' or extreme value which draws the mean away from the true centre. Neither of the last two datasets is effectively represented by its mean.

Dataset A: 5.2 4.3 2.9 4.6 6.2 6.9 4.7 4.5 6.0 4.7
Dataset B: 2.4 5.1 8.5 7.0 6.1 3.9 3.2 1.4 9.4 3.0
Dataset C: 2.5 1.2 2.2 1.4 1.4 8.7 8.1 7.7 7.8 9.0
Dataset D: 3.0 2.3 3.2 3.6 4.0 3.1 3.3 4.5 3.0 20.0

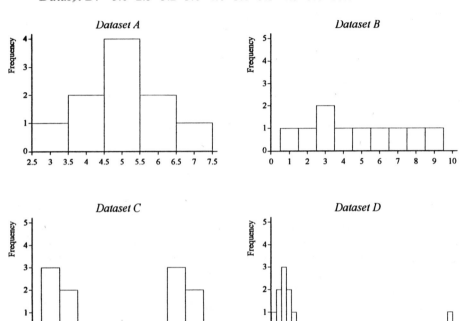

Fig 2.19 Histograms of the four datasets with mean 5.

The median

Another way of defining representative points or averages for sets of data is the **median** which has half the dataset values below it and half above it. In terms of the histogram, half the area will be below the median and half above it. Of course, if the histogram happens to be symmetrical about a vertical line, that line will be both the mean and the median.

To find the median of a set of data, first arrange the values in ascending order. If there is an odd number of values, the median is the middle value of the ordered sample. If there is an even number of points, any value between the middle two points would have the required property. In this case we define the median to be half way between them (i.e. the mean of the middle two points).

The stem-and-leaf diagram may be used to find the median. The cumulative count, as recorded in the Minitab output of Fig 2.7, identifies the row containing the median and it is a simple matter to count along that row to find the value or values required. For a dataset of 118 values, the median is the average of the 59th and 60th ordered values. Our stem-and-leaf diagram shows that there are 45 values before the row containing the median, so we need to average the 14th and 15th values in that row (both of which are 8).

Example 2

Calculate the means and medians of *chola* and *cholb* (Dataset 1). Relate your results to the skewness revealed by Fig 2.17.

```
MTB > mean 'chola'
   MEAN      =       6.7281
MTB > median 'chola'M
   MEDIAN =          6.8550
```

```
MTB > mean 'cholb'
   MEAN      =       7.3525
MTB > median 'cholb'
   MEDIAN =          7.2900
```

Fig 2.20 Minitab calculations of means and medians.

From the Minitab output (Fig 2.20) the mean of *chola* is less than its median (skewed to the left), whereas the mean of *cholb* is greater than its median (skewed to the right). Skewness is thus best understood with reference to a histogram, but may be adequately described by reference to statistics such as mean and median.

Modes

A value is a **mode** of a set of data if it occurs at least as frequently as any other value in the set. Where the data are drawn from a discrete set of values, such as integers, equal observations really are *exactly* equal, and the modes are the most common values. For data from a continuous set of values, equal observations are only equal to the level of accuracy employed in collecting the data. Thus modes are dependent upon this level of accuracy. For such data it is useful to produce a frequency table, and define the modes to be the most commonly observed classes. In a histogram the tallest rectangles represent the modes.

Example 3

A football team has scored the following number of goals in the last 20 matches:

5 0 1 1 0 4 2 4 0 1 3 2 1 1 4 0 1 3 1 2

Given no information about their opponents, what is the most likely number of goals that they will score in their next match?

The most likely result corresponds in the sample to the most common value (i.e. the mode of the data) which is the value 1. This is immediately obvious if a frequency table or histogram is drawn of this data.

Example 4

What is the mode of the Irish population data (Dataset 3)?

From the histogram in Fig 2.3, it is clear that the mode is the category of 5–10 year olds.

Percentiles and quantiles

Not all measures of location are targeted on the centre of the distribution. The median is, in fact, a special case of a whole class of measures of location called **percentiles** or **quantiles**. For example, the 30th percentile (also called quantile 0.3) has 30% of the values below it. The median is the 50th percentile (quantile 0.5).

To find quantile q, in a dataset of n values, we must first calculate an index value, $i = nq + 0.5$ and then use i as an index for the ordered dataset.

Example 5

Find quantile 0.4 (the 40th percentile) of the following set of data:

$$5\ 7\ 2\ 10\ 9\ 8\ 7\ 3\ 2\ 10\ 15\ 5\ 8$$

Here $n = 13$, so $i = 13 \times 0.4 + 0.5 = 5.7$. So the required quantile is the '5.7th' value in the ordered dataset: 2 2 3 5 5 7 7 8 8 9 10 10 15. The 5th value is 5 and the 6th value is 7, so we must find a value 0.7 of the way from 5 to 7. This is achieved by **linear interpolation**: $(1 - 0.7) \times 5 + 0.7 \times 7 = 6.4$. So the value 6.4 is quantile 0.4.

Example 6

In Dataset 1, how high must a person's *cholb* value be to place them in the top 10%?

To answer this we need to find the 90th percentile. The index value is

$$i = 118 \times 0.9 + 0.5 = 106.7$$

The Minitab session in Fig 2.21 sorts the values into ascending order, and then performs the linear interpolation. The result is 8.432, so any person with a *cholb* value above this is in the top 10% 'at risk' category.

```
MTB > sort 'cholb' c12
MTB > let k3=(0.3*c12(106))+(0.7*c12(107))
MTB > print k3
K3         8.43200
```

Fig 2.21 Minitab Session to find the 90th percentile of *cholb*.

Quartiles and hinges

The 25th and 75th percentiles are called the **lower quartile** and the **upper quartile**, respectively. Together with the median, they divide the dataset into four equal portions.

Example 7

Find the median and both quartiles of the data in Example 5.

Solving the quantile index equation for quantiles 0.25, 0.5 and 0.75:

$i_1 = 13 \times 0.25 + 0.5 = 3.75$
$i_2 = 13 \times 0.5 + 0.5 = 7$
$i_3 = 13 \times 0.75 + 0.5 = 10.25$

The lower quartile is the 3.75th value: 0.75 of the way from 3 to 5 = 4.5.
The median is the 7th value: 7.
The upper quartile is the 10.25th value: 0.25 of the way from 9 to 10 = 9.25.
The upper quartile in Example 7 demonstrates that, when the original data comprises whole numbers, the quartiles may involve two decimal places (.25 or .75).

An alternative method of dividing the data into approximately equal quarters employs **hinges**. The **lower hinge** is the median of that portion of the data which is less than or equal to the median of the whole dataset. The **upper hinge** is the median of that portion of the data which is greater than or equal to the median of the whole dataset.

Example 8

Find the upper and lower hinges of the data in Example 5.

We have seen that the median of this data is 7.
The portion of the data which is less than or equal to the median is: 2 2 3 5 5 7 7.
The median of this portion (the lower hinge) is 5.
The portion of the data which is greater than or equal to the median is: 7 7 8 8 9 10 10 15.
The median of this portion (the upper hinge) is 8.5.

The five-number summary

The median and the upper and lower quartiles, together with the largest and smallest values in the dataset (the **maximum** and **minimum**), form a standard summary which helps us to compare the salient features of datasets. These five values are called the **five-number summary**, and form the basis of the **boxplot** described in Section 2.8.

EXERCISES ON 2.6

1. Using your stem-and-leaf display of the blood-pressure data for Exercise 1 in Section 2.4, or otherwise, calculate the three quartiles. Can you give a reason why such data may be skewed to the right?
2. Calculate the three quartiles of each of the datasets illustrated in Fig 2.19.

2.7 Statistics that measure dispersion or spread

When you are trying to summarise the properties of a dataset, statistics concerned with location are very important, but for many purposes measures of the **spread** (or dispersion) of the data are of equal importance. As with location, many measures of spread have been devised.

The range of the data

When asked the range of the *cholb* data, you may be inclined to reply 'it ranges from 4.43 to 9.73'. These, of course, are the minimum and maximum values – two measures of location from the five-number summary. What statisticians call the **range** is the difference between these two values: for *cholb*, 5.3. This definition of the range is a simple measure of spread.

The variance and the standard deviation

The **variance** (symbol s^2) is an average of the squares of the deviations from the mean:

$$s^2 = \frac{1}{n-1} \sum_{i=1}^{n} (x_i - \bar{x})^2$$

It is suitable for use with interval/ratio data. Unlike the range, the units in which variance is measured are not the units of the original data. For example, if you measure the data in metres, the variance will be in square metres. We use the divisor $n - 1$ because we can show mathematically that it makes the result right 'on average'. Some statisticians prefer to define the variance with n as the divisor. This is a perfectly acceptable definition, but formulae for statistics involving s, such as t-statistics discussed in Chapter 5, must be modified if you use this definition. **Consistency is essential** – all formulae in this book assume that you are using $n - 1$.

The **standard deviation** (symbol s) is the positive square root of the variance. It is one of the most frequently used measures of spread, and is expressed in the same units as the original data.

Example 9

Twenty students record their weights in kg, creating the following dataset:

49 50 60 74 61 72 63 76 68 58 68 73 60 69 63 58 72 57 64 85

Calculate the mean, variance and standard deviation.

Adding the values and dividing the result by 20 gives $\bar{x} = 65$.
  Subtracting this mean from each observation gives:

−16 −15 −5 9 −4 7 −2 11 3 −7 3 8 −5 4 −2 −7 7 −8 −1 20

Squaring each of these values:

256 225 25 81 16 49 4 121 9 49 9 64 25 16 4 49 49 64 1 400

The sum of these squares is 1516, leading to the variance, $s^2 = \frac{1}{19} 1516 = 79.789$ and standard deviation, $s = 8.932$.

Measures of spread based on the quartiles

A variety of measures of spread may be defined in terms of the quartiles. They are suitable for ordinal or ranked data as well as interval/ratio data. The **interquartile range** (also called the **mid-spread**) is the difference between the upper and lower quartiles. The **semi-interquartile range** (also called the **quartile deviation**) is the interquartile range divided by 2. For data which are normally distributed (see Chapter 3) the semi-interquartile range, Q, is related to the standard deviation, s, by $Q = 0.6745s$. For such data, $4Q$ on either side of the median may be expected to cover all values. This fact leads to the definition of 'inner fence' for a diagram called the boxplot.

> John Wilder Tukey was born in New Bedford, Massachusetts, on 16 June 1915, and made his career as a statistician at Princeton and at Bell Telephone Laboratories. In the 1950s he developed the jack-knife technique, an important tool used by statisticians to improve estimation precision. From the 1960s onwards Tukey pioneered exploratory data analysis, developing the stem-and-leaf plot and the boxplot, among many other EDA displays and methods of analysis. His book *Exploratory Data Analysis* (Addison-Wesley, 1977) is still the most widely quoted authority on the subject.

2.8 Boxplots

A **boxplot** (also called a **box-and-whisker plot**) is a diagram designed by John Tukey as a pictorial representation of the five-number summary and to identify extreme values. It is commonly used as an EDA tool. Figure 2.22 shows a boxplot for the *cholb* data.

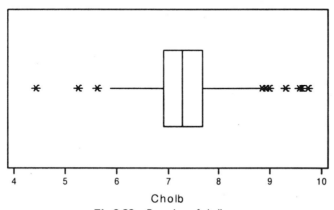

Fig 2.22 Boxplot of *cholb*.

The 'box' covers the interval between the lower and upper quartiles, and the bar shows the position of the median. The 'whiskers' extend approximately a further three semi-interquartile ranges beyond each quartile (in fact they end at the last actual data values inside that range) and thus cover the values that conform to the expectations of a normal distribution. Values outside this range are plotted individually and identified as 'extreme'. Those lying outside a further three semi-interquartile ranges on either side are called 'very extreme'.

Stages in the production of the boxplot of *cholb*

- Calculate the median, the lower and upper quartiles and the semi-interquartile range:

    Median = 7.29   Lower quartile = 6.92   Upper quartile = 7.66

    Semi-interquartile range = 0.5(7.66 − 6.92) = 0.37.

- Calculate the inner fences – three semi-interquartile ranges beyond the quartiles:

    Lower inner fence = 6.92 − 3 × 0.37 = 5.81

    Upper inner fence = 7.66 + 3 × 0.37 = 8.77

- Identify the adjacent values (actual data points inside and nearest to the fences):

    Lower adjacent value = 5.9   Upper adjacent value = 8.76

- Calculate the outer fences – six semi-interquartile ranges beyond the quartiles:

    Lower outer fence = 6.92 − 6 × 0.37 = 4.7

    Upper outer fence = 7.66 + 6 × 0.37 = 9.88

- Identify extreme and very extreme values:

    Extreme values below the lower inner fence: 5.25, 5.62
    Extreme values above the upper inner fence: 8.82, 8.89, 8.97, 9.29, 9.55, 9.6, 9.73
    Very extreme value below the lower outer fence: 4.43

The boxplot is a powerful tool for comparing distributions. Figure 2.23 shows four boxplots: *cholb* for men, *cholb* for women, *chola* for men, and *chola* for women.

Like back-to-back and multiple histograms/stem-and-leaf diagrams, the multiple boxplots are suitable for comparing distributions when all datasets are measured in the same units.

EXERCISES ON 2.8

1. Produce six simultaneous boxplots of men's and women's gold, silver and bronze medal times for the 4 × 100 metres (see Dataset 2, Appendix A).
2. Draw a boxplot of the blood-pressure data of Exercise 1 in Section 2.4. Explain how the skewness of this data can be deduced from the boxplot.

2.9 Paired values

When you want to explore the relationship between two datasets in the form of paired values, possibly measured in different units, you have a large variety of statistical techniques from which to choose. As with single datasets, if you suspect

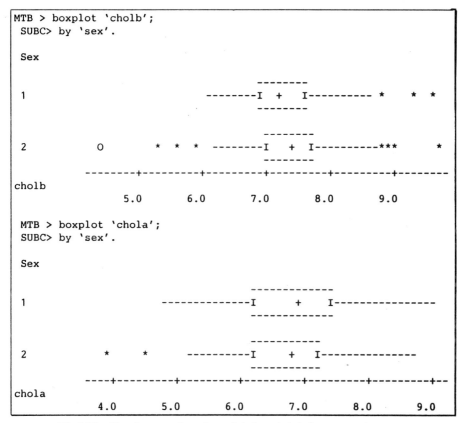

Fig 2.23 Simultaneous boxplots of *cholb* and *chola* for men and women.

a particular relationship (e.g. a linear one) you can choose appropriate statistics and displays; otherwise you can adopt the standard EDA approach.

Scatter plots

The simplest EDA tool for paired data is the **scatter plot**, in which we plot one variable on the horizontal axis and the other on the vertical axis, representing each 'case' (data item) by a point or a symbol. For example, we can plot *chola* against *cholb* (Fig 2.24).

What insight does this representation give us that cannot be derived from the simultaneous boxplots? The main advantage is that we can now see what has happened to individuals. We can note, for example, that one individual has reduced cholesterol level from nearly 9 to below 5 (case 117: *cholb* = 8.89, *chola* = 4.51), while another has increased from below 7 to nearly 9 (case 80: *cholb* = 6.71, *chola* = 8.99).

If we plotted the line *chola* = *cholb* on this scatter plot, the points above the line would be those whose cholesterol level has increased (30 cases, plus one exactly on the line).

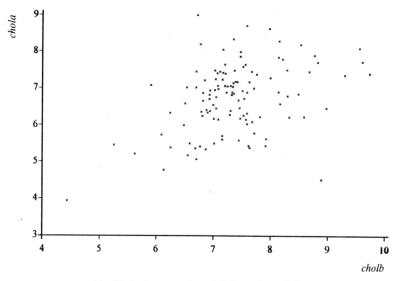

Fig 2.24 Scatter plot of *chola* against *cholb*.

Is there any relationship between *chola* and *cholb*? Generally, those with high *cholb* also have high *chola* and vice versa, but there is a great deal of scatter. We can investigate the *change* in cholesterol level for individual cases by plotting *cholinc* (= *chola* − *cholb*) against *cholb* (Fig 2.25).

Once again, there is a lot of scatter, but it is also reasonably clear that people with higher *cholb* tend to experience the greatest reduction – in fact, virtually everyone with a *cholb* over 8 reduced their cholesterol level.

A simple, but sometimes very powerful, extension to the scatter plot involves *labelling* the points according to some categorical variable. For example, we can

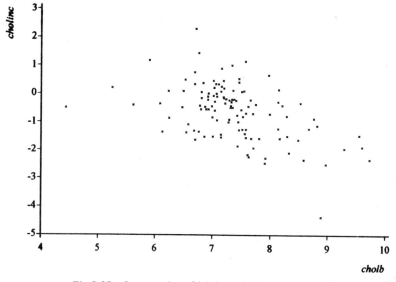

Fig 2.25 Scatter plot of (*chola* − *cholb*) against *cholb*.

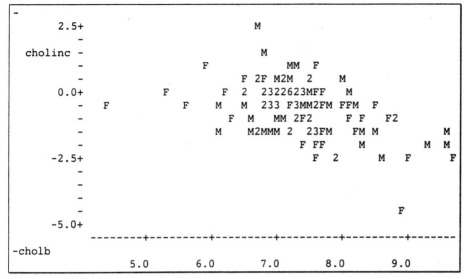

Fig 2.26 Scatter plot of (*chola* − *cholb*) against *cholb* labelled by *sex*.

investigate the relationship between (*chola* − *cholb*) and *cholb* according to *sex* (Fig 2.26). Using Minitab, we can get this sort of scatter plot using the LPLOT command.

It is immediately apparent that all the low *cholb* cases are female and of the six largest *cholb* cases, the three female ones all achieved greater reductions than the three male ones. The two greatest increases were also male.

A scatter plot of *times* against *year* for the Olympics data (Dataset 2, Appendix A) by *sex* reveals a different sort of information (Fig 2.27).

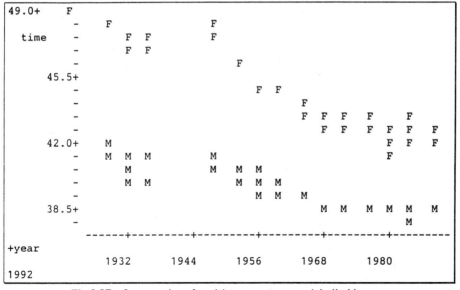

Fig 2.27 Scatter plot of *medal times* against *year* labelled by *sex*.

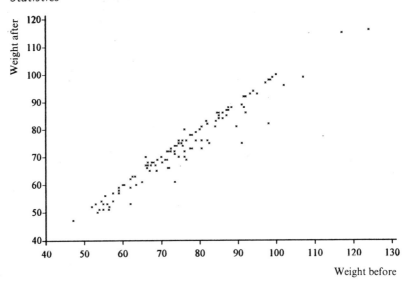

Fig 2.28 Scatter plot of *wta* against *wtb*.

An unlabelled scatter plot immediately reveals the two separate lines. The labelling explains the difference. Also apparent is the absence of data from the war years.

Sometimes it is fairly obvious that an underlying linear (or very nearly linear) relationship is present. Suppose we plot the two weights, *wta* against *wtb* (Fig 2.28).

As we may have expected, there is a strong relationship. In this case, we can make a good 'guess' at the sort of straight line that may be the underlying model for the data. We may even be able to predict values of *wta* from corresponding values of *wtb* with some degree of accuracy.

2.10 Correlation

It is clear from the scatter plots in Figs 2.24–2.28 that there is a weak relationship between some pairs of variables, such as *chola* and *cholb*, and a strong relationship between others, such as *wta* and *wtb*. Statistics have been designed to measure the strength of this relationship, or **correlation**.

The Pearson correlation coefficient

The most commonly used measure of correlation between pairs of values (x_i, y_i) is the product–moment correlation coefficient of Karl Pearson (1857–1937):

$$r = \frac{S_{xy}}{\sqrt{S_{xx} S_{yy}}}$$

where

$$S_{xx} = \sum_{i=1}^{n}(x_i - \bar{x})^2 \quad S_{yy} = \sum_{i=1}^{n}(y_i - \bar{y})^2 \quad S_{xy} = \sum_{i=1}^{n}(x_i - \bar{x})(y_i - \bar{y})$$

Exploring and Presenting Data 33

r can take values between -1 and $+1$. The two extremes represent a perfect linear relationship, with every point in the scatter plot lying *exactly* on a straight line. If the y values tend to increase as x increases, r will be positive; if the y values decrease as x increases, r will be negative. $r = 0$ means that there is no *linear* relationship between y and x. The value of r does not depend on the units in which the variables are measured.

> The Pearson correlation coefficient was first proposed in 1877 by Francis Galton (1822–1911), a British psychologist who was a cousin of Charles Darwin. He received a gold medal from the Royal Geographical Society for exploring South West Africa. His work *Hereditary Genius* (1869) argued that mental characteristics can be inherited, and he went on to study the behaviour of twins, attempting to distinguish genetic and environmental influences. Karl Pearson (1857–1936), a British statistician, was Galton's student. He clarified Galton's work on correlation and established the technique of multiple regression. In 1900, Pearson devised the chi-squared goodness-of-fit test, which has a fair claim to being the most commonly applied statistical procedure ever invented (see Chapter 5).

Example 10

A farmer varies the amount of fertilizer used in different fields in an attempt to discover the relationship between the weight of fertilizer per acre, X, and the crop yield per acre, Y. Over a period of time he records his results as pairs (X, Y):

Fertilizer, X	7	5	5	10	9	3	6	12	11	12
Yield, Y	9	6	5	9	11	5	7	12	13	13

Produce a scatter plot and calculate the correlation coefficient.

The scatter plot is shown in Fig 2.29.

$$\sum_{i=1}^{10} x_i = 80 \text{ so } \bar{x} = 8 \qquad \sum_{i=1}^{10} y_i = 90 \text{ so } \bar{y} = 9$$

Table 2.1 shows the calculation of the sums of squares required for r.
From this table,

$$r = \frac{87}{\sqrt{94 \times 90}} = 0.946$$

We can easily confirm this result in Minitab:

```
MTB > corr 'fertzer' 'yield'

Correlation of fertzer and yield = 0.946
```

34 Statistics

```
MTB > name c1 'fertzer'
MTB > set into 'fertzer'
DATA> 7 5 5 10 9 3 6 12 11 12
DATA> end
MTB > name c2 'yield'
MTB > set into 'yield'
DATA> 9 6 5 9 11 5 7 12 13 13
DATA> end
MTB > plot 'yield' against 'fertzer'
```

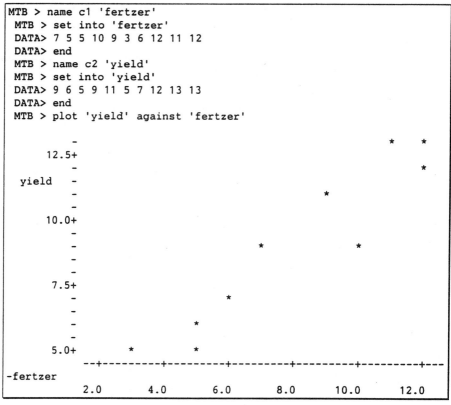

Fig 2.29 Minitab output for the scatter plot.

Table 2.1 The calculation of the sums of squares for r.

x_i	y_i	$x_i - \bar{x}$	$(x_i - \bar{x})^2$	$y_i - \bar{y}$	$(y_i - \bar{y})^2$	$(x_i - \bar{x})(y_i - \bar{y})$
7	9	−1	1	0	0	0
5	6	−3	9	−3	9	9
5	5	−3	9	−4	16	12
10	9	2	4	0	0	0
9	11	1	1	2	4	2
3	5	−5	25	−4	16	20
6	7	−2	4	−2	4	4
12	12	4	16	3	9	12
11	13	3	9	4	16	12
12	13	4	16	4	16	16
80	90	0	$S_{xx} = 94$	0	$S_{yy} = 90$	$S_{xy} = 87$

This value is quite near to +1, i.e. there is a high **positive** correlation between *fertilizer* and *yield*. In other words, the points lie close to a straight line, and when *fertilizer* increases *yield* tends to increase.

For comparison, the correlations for the scatter plots in Figs 2.24, 2.25 and 2.28 are shown in Table 2.2.

Table 2.2 Correlations for Figs 2.24, 2.25 and 2.28.

Variable 1	Variable 2	Correlation
chola	cholb	0.390
chola − cholb	cholb	−0.477
wta	wtb	0.972

The SD line

What is the line around which the points are clustered when they are highly correlated? It is sometimes called the **SD line**. It passes through the centre of gravity of the scatter plot, $(\bar{x}, \bar{y})$, and it passes through points which are an equal number of standard deviations away from $(\bar{x}, \bar{y})$ for both variables.

We can express all linear relationships in the form $Y = \beta_0 + \beta_1 X$ for some values of β_0 and β_1. When this line is plotted on a graph, β_0 is the **intercept**, i.e. the value of Y obtained when $X = 0$, and β_1 is the **gradient**, i.e. the amount by which Y increases when X is increased by one unit.

The gradient of the SD line is $\sqrt{S_{yy}/S_{xx}}$ for positive correlation, and $-\sqrt{S_{yy}/S_{xx}}$ for negative correlation.

For the data in Example 10, the SD line has the equation $y = 1.172 + 0.978x$. This line is superimposed on the scatter plot of the data in Fig 2.30, together with another line, called the regression line, discussed in Section 2.11.

EXERCISE ON 2.10

1. Take a systematic sample from Dataset 1, by considering every tenth case (i.e. patient 10, patient 20, patient 30, ...). You should end up with 11 cases. Calculate the correlations between (a) *chola* and *cholb*, (b) *chola − cholb* and *cholb*, (c) *wta* and *wtb*. Compare your results with those in Table 2.2 for the whole dataset.

2.11 Regression analysis

The process by which we arrive at a suitable straight line or curve to predict values of one variable from the values of others is called **regression analysis**. The SD line is one of many candidates for this purpose. Another way to do it is to take a ruler, place it on a scatter plot such as Fig 2.28, and draw a line through the middle of the data. The line you get will have several desirable properties – for example, you will probably have ignored any obviously extreme values which do not conform to the general model, but another person would take the ruler, draw another line, and claim that it is 'better'. An important point to note is that the line you choose may

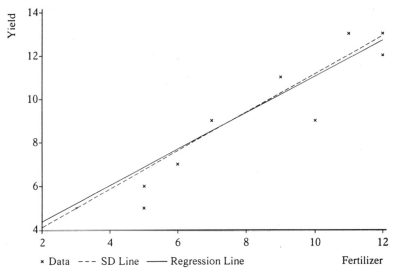

Fig 2.30 Scatter plot of fertilizer data, with SD line and regression line.

depend on whether you are trying to predict the y values from the x values or vice versa. How can we devise an objective criterion by which a line can be selected?

Agreement amongst statisticians on a 'best' line is no more possible than politicians agreeing on a 'best' policy or university students agreeing on 'a good lecturer'! What is therefore important is to propose statistical criteria which imply a 'best line' – whilst the applicability of the criteria may be (and should be) questioned by anyone and everyone, their acceptance leads to a definitive line.

Simple linear regression

We shall confine our attention to the two-variable problem (called **simple**) where the relationship is a straight line (**linear**).

In regression the two variables have distinct roles. The one we are trying to predict is called the **response variable**, and the one from which we are deriving the predictions is called the **explanatory variable**. In non-simple models there will be more than one explanatory variable, and in multivariate models, there may be more than one response variable. For explanatory variable X and response Y, we must find a line of the form $E(Y) = \beta_0 + \beta_1 x$ that predicts the expected response value when the explanatory variable $X = x$.

For reasons discussed in detail in Chapter 4, a technique call the **method of least squares** gives a good solution to this problem, and provides **estimates** $\hat{\beta}_0$ and $\hat{\beta}_1$ for the parameters β_0 and β_1.

The least-squares line estimates the gradient β_1 with the value $\hat{\beta}_1 = S_{xy}/S_{xx}$ and the intercept β_0 with the value $\hat{\beta}_0 = \bar{y} - \hat{\beta}_1 \bar{x}$, where S_{xx} and S_{xy} are defined as in Section 2.10. Like the SD line, the regression line passes through the point $(\bar{x}, \bar{y})$.

For the *fertilizer* data of Example 10, note that we can read the values of S_{xx} and S_{xy} from the totals of the table: $S_{xx} = 94$ and $S_{xy} = 87$. So

$$\hat{\beta}_1 = 87/94 = 0.9225 \quad \text{and} \quad \hat{\beta}_0 = 9 - 0.9255 \times 8 = 1.596.$$

```
MTB > regress 'yield' on 1 predictor 'fertzer'

The regression equation is
yield = 1.60 + 0.926 fertzer

Predictor          Coef         Stdev        t-ratio          p
Constant         1.5957        0.9619          1.66       0.136
fertzer          0.9255        0.1123          8.24       0.000

s = 1.089       R-sq = 89.5%       R-sq(adj) = 88.2%

Analysis of Variance

SOURCE            DF           SS           MS           F         p
Regression         1       80.521       80.521       67.96     0.000
Error              8        9.479        1.185
Total              9       90.000
```

Fig 2.31 Minitab output for regression calculations.

The least-squares regression line is therefore $y = 1.596 + 0.9255x$. Note that this is a different line from the SD line, though in this case, they are quite similar, because r is near 1.

The values $\hat{\beta}_0$ and $\hat{\beta}_1$ calculated in this way are called the **sample regression coefficients**. They may be calculated in Minitab, together with a great deal of other information, using the REGRESS command, as shown in Fig 2.31. This gives the regression line, followed by further details of the regression coefficients, which we shall explore in Chapters 4 and 5.

The regression line is superimposed on the scatter plot of the fertilizer data in Fig 2.30. From this we can read off predictions of yield, given the amounts of fertilizer applied. For example, with a fertilizer value of 4, we read off a yield of a little over 5, and for a fertilizer value of 8, the yield is about 9. Going back to the equation of the regression line, we can check these results:

$$1.596 + 0.9255 \times 4 = 5.298 \quad \text{and} \quad 1.596 + 0.9255 \times 8 = 9.000.$$

Summary

- The overall shape of a set of data may be summarised numerically in a *frequency distribution* and displayed pictorially in a *histogram*, a *pie chart*, a *stem-and-leaf display*, or a *boxplot*.
- Various multiple and back-to-back versions of histograms, pie charts, stem-and-leaf displays and boxplots facilitate comparison of two or more datasets (or one dataset split by a categorical variable). For meaningful comparisons, datasets should be measured in similar units.
- The *mean, median* and *mode* are statistics which describe the *location* of the centre of a distribution.
- *Percentiles, quantiles, quartiles* and *hinges* are statistics which describe the shape of a distribution.

38 Statistics

- The *five-number summary* comprises the maximum and minimum, upper and lower quartiles and median.
- The *variance, standard deviation, interquartile range* and *semi-interquartile range* are measures of the *spread* of a distribution.
- *Scatter plots* display the relationship between two variables. They do not have to be measured in similar units. *Labelled scatter plots* allow splitting by categorical variables.
- The *Pearson correlation coefficient*, r, measures the linear relationship between two variables. r takes values between -1 and 1, the two extremes representing a perfect linear relationship, and 0 representing no linear relationship.
- The *SD line* passes through $(\bar{x}, \bar{y})$ and points which are an equal number of standard deviations away from $(\bar{x}, \bar{y})$ for both variables.
- *Regression analysis* enables us to predict values of one variable from the values of other variables. The linear regression of y on x gives the regression line $y = \beta_0 + \beta_1 x$ where the coefficients are estimated by $\hat{\beta}_1 = S_{xy}/S_{xx}$ and $\hat{\beta}_0 = \bar{y} - \hat{\beta}_1 \bar{x}$. The regression line also passes through the point $(\bar{x}, \bar{y})$.

FURTHER EXERCISES

1. The data below gives recordings of air pollution on consecutive days in a city in the United States. Using your calculator, calculate (a) the sample mean, (b) the sample standard deviation. Construct an ordered stem-and-leaf plot. Comment on the data in the light of your results.

   6.6 2.1 4.5 4.9 4.9 2.3 4.8 2.9 4.0 4.7 8.3 5.3
   4.8 4.4 3.4 5.6 4.8 2.9 1.9 2.4 3.3 5.3 5.2 4.6
   6.5 3.9 5.3 8.1 2.3 2.0 4.6 4.5 3.1 3.6 6.3 4.3

2. The following data is the time between failures of a computer system. Present a short report discussing the major features of the data.

817	357	7	2	64	84	545	86	109	141
33	367	28	90	325	13	748	6	808	300
227	130	66	113	233	8	3	137	425	330
55	1149	82	60	19	301	1	45	671	31
145	76	378	532	115	194	193	290	24	8
371	232	1350	174	180	1224	121	23	112	39
457	13	111	6	15	48	237	600	6	41

3. Draw simultaneous boxplots of the drop in cholesterol level (*cholb* − *chola*) for the four groups defined by the categorical variables *sex* and *health visitor/dietician*.

TUTORIAL PROBLEM

A good reference library should stock a copy of the *Abstract of British Historical Statistics*. Locate in that source (or equivalent source) a table giving the number of live births (male and female) in England and Wales in the years 1920–1925. Investigate the data and draw up a short report (including appropriate statistical diagrams) of your findings. The following questions may help you.

1. Describe the behaviour of the total number of live births during that period.
2. Look at the sex ratio (male to total) distribution for the 26 years as a sample.
3. Look at the sex ratio and how it behaves over time.
4. See if there is anything strange happening which might make you question your calculations **or** the source material.

3 • Probability Distributions

3.1 Probability, random variables and expected values

The managers of a university computer laboratory of 25 PCs keep a record of the number of machines that malfunction each day, and produced the following frequency table after 100 days:

Number of malfunctions	0	1	2	3	4
Frequency	36	34	21	7	2

Can they use this data to make predictions about the future behaviour of the machines?

The answer is yes, but they will have to make some assumptions about the way the future behaviour is related to the past behaviour. For example, will each machine have the same propensity to fail in future as in the past – or is there some 'ageing effect', i.e. will more machines fail as they get older? If we assume the same proportions, then simple predictions can follow.

On how many days in a 365 day year would you expect more than two machines to fail?

A simple approach to this problem is to note that more than two machines failed on nine days out of 100 in the observed data (add together the frequencies of 3 and 4). So on 9% of days (proportion 0.09) we may expect more than two failures; 9% of 365 is 32.85, so we can predict that more than two machines will fail on about 33 days in a 365 day year.

The proportion of occasions on which a particular event may be expected to take place is the **probability** of that event. So the probability that more than two machines fail on a given day is *estimated* to be 0.09. Of course, if the managers collected a new set of data, they would obtain a new (and maybe different) estimate of the probability. Furthermore, the method used here is only one of several methods that we could consider using to estimate the probability.

When the results of a 'statistical experiment' have numerical values attached to them, we call the resulting quantity a **random variable**. The number of PCs which malfunction on a given day is a random variable. Here the statistical experiment consists of running the laboratory for a day and counting the number of machines that fail. We usually refer to random variables by means of capital letters towards the end of the alphabet, e.g. let the number of machines failing be X.

What sort of values can X take? In this experiment, the only possible values are the integers from 0 to 25. A random variable which takes a countable set of values is called a **discrete** random variable, and we describe its probability structure by means of a **probability distribution** (pd), which is a set of pairs of quantities – each possible value paired with its corresponding probability.

So what is the probability distribution of X? The short answer is that we don't really know. We know the possible values of X, but we can only *estimate* the corresponding probabilities. Using the same technique as before, we can construct the following *empirical* probability distribution:

x	0	1	2	3	4	5
$P(X = x)$	0.36	0.34	0.21	0.07	0.02	0

Note that we use a lower case letter to represent the possible values of the random variable, and the expression $P(X = x)$ as shorthand for 'the probability that the random variable X takes the value x'. The probabilities of all values above 5 are also *estimated* to be zero. This is a major drawback of the empirical probability distribution – we know that it is possible for X to be 5, and yet we estimate the probability as zero.

On average, how many machines do we *expect* to fail on a given day?

Again, going back to the collected data, a total of 105 failures were recorded in 100 days – obtained from $(1 \times 34) + (2 \times 21) + (3 \times 7) + (4 \times 2)$. So we expect on average 1.05 machines to fail on a given day. This is the **mean** or **expected value** of the random variable X, usually written $E(X)$. It is easily calculated from the probability function by multiplying each possible value by its corresponding probability and summing the results:

$$E(X) = \sum_{i=1}^{n} x_i p_i$$

where x_i are the values and p_i are the corresponding probabilities,

$$E(X) = (0 \times 0.36) + (1 \times 0.34) + (2 \times 0.21) + (3 \times 0.07) + (4 \times 0.02) = 1.05$$

Note, however, that this is once again an *estimate* based on the estimated probability function.

How much will X vary about this mean or expected value? We can measure variability by calculating the **variance**, $Var(X)$. This is defined as

$$Var(X) = E[X - E(X)]^2$$

or, equivalently,

$$Var(X) = E(X^2) - [E(X)]^2$$

The expected value of a function such as X^2 is found by multiplying each possible value of the function by its corresponding probability and summing the results:

$$E(X^2) = (0^2 \times 0.36) + (1^2 \times 0.34) + (2^2 \times 0.21) + (3^2 \times 0.07) + (4^2 \times 0.02)$$
$$= 2.13$$

So
$$Var(X) = 2.13 - (1.05)^2 = 1.0275$$

On how many days in a five-year period would you expect five or more machines to fail?

We have estimated the probability of five or more to be zero, and so the obvious answer is that the event will never happen. But common sense tells us that it probably will. We need a more sophisticated method of estimating the probability function of X.

EXERCISES ON 3.1

1. A man aiming at a target receives 10 points if his shot is within 1 cm of the centre of the target, five points if it is between 1 and 3 cm, and three points if it is between 3 and 5 cm. Find the expected number of points scored if the man's shot is uniformly distributed in a circle of radius 8 cm.
2. An ambulance travels back and forth at a constant speed along a road of length L. At a certain moment of time an accident occurs at a point on the road. Making suitable assumptions about the position of the ambulance X and the position of the accident Y determine the distribution of the distance the ambulance has to travel, and the expected value of this distribution.

   Hint: Divide the road length L into a fixed number (say 10) equal segments, and consider the possible values of X and Y with respect to these segments.

3.2 The binomial distribution

If we make some further assumptions about the probability structure of X, we may be able to infer something about the probabilities of obtaining values larger than 4. The form of this particular experiment resembles a basic statistical structure called **Bernoulli trials**. The requirements for Bernoulli trials are:

1. Each trial has two possible outcomes, usually referred to as 'success' and 'failure'.
2. The outcomes are statistically independent, i.e. knowing the result of one trial has no influence over the results of any other trials.
3. The probability of a 'success' is the same for each trial.

If we count a malfunctioning machine as a 'success', then clearly requirement 1 is satisfied. Statistical independence is more difficult. If the machines are networked, and any of the malfunctions are due to network problems which may affect more than one machine, then this requirement will be violated. Similarly, if there are problems with the power supply to several machines, the trials will not be independent. But if these problems can be ruled out, requirement 2 may well hold. Finally, each machine must have the same probability of failing, i.e. we must not have some machines that are more prone to failure than others. If all three hold, we have Bernoulli trials.

If the number of trials is fixed, i.e. not dependent upon the outcome of the experiment, then the number of successes X observed in n trials with probability of success p will have a **binomial distribution** with parameters n and p, denoted by $Bi(n, p)$. The probabilities associated with all possible outcomes of a binomial distribution can be calculated, provided n and p are known, using the formula

$$P(X = x_i) = \binom{n}{x_i} p^{x_i}(1-p)^{n-x_i} \quad (x_i = 0, 1, 2, \ldots, n)$$

where

$$\binom{n}{x} = \frac{n!}{(n-x)!\, x!}$$

(On some calculators, this function is written ^nC_x.)

These probabilities can also be found in books of tables, or calculated by a computer package. With the extra assumptions now made, we can use the binomial distribution to estimate the pd of X. Each day 25 machines could fail, so the number of trials n is 25. The probability of a malfunction, p, must be estimated from the data. The total number of malfunctions is 105 out of a possible 2500, giving a proportion 0.042, so this is our best estimate of p.

We can calculate the resulting pd using Minitab or the formula (see Fig 3.1).

The first thing to note is that these values are not greatly different from the empirical estimates. In other words, the binomial distribution appears to fit the observed values reasonably well. You should always consider this whenever you use a distribution to make predictions based on observed data. The usual procedure is to derive 'expected' frequencies for each category, by multiplying each probability by the total number of observations (in this case 100) as in Fig 3.2.

There is very little evidence of discrepancy here between the observed and expected frequencies. We will consider a formal test for such a discrepancy in Chapter 5.

Now return to the problem of estimating $P(X \geq 5)$. This is carried out by Minitab in Fig 3.3.

So $P(X \geq 5) = 0.0034306$.

In five years (1825 days) we would expect $1825 \times 0.0034 = 6.205$, i.e. about six days when five or more machines malfunction. Assuming that our use of the binomial distribution is valid, we have managed to attach a probability to an event that was not actually observed in the data.

```
MTB > pdf;
SUBC> binomial n=25, p=0.042.

BINOMIAL WITH N =   25   P =  0.042000
     K              P( X = K)
     0              0.3421
     1              0.3749
     2              0.1973
     3              0.0663
     4              0.0160
     5              0.0029
     6              0.0004
     7              0.0001
     8              0.0000
```

Using the formulae:

$P(X = 0) = (1-p)^{25}$

$P(X = 1) = 25p(1-p)^{24}$

$P(X = 2) = \dfrac{25 \times 24}{1 \times 2} p^2 (1-p)^{23}$

$P(X = 3) = \dfrac{25 \times 24 \times 23}{1 \times 2 \times 3} p^3 (1-p)^{22}$

and so on.

Fig 3.1 Calculation of the binomial pd.

```
MTB > set c1
DATA> 36 34 21 7 2
DATA> end
MTB > set c2
DATA> 0:8
DATA> end
MTB > pdf probabilities of c2 in c3;
SUBC> binomial n=25, p=0.042.
MTB > let c4=round(c3*100)
MTB > name c1 'observed'
MTB > name c2 'values'
MTB > name c3 'probs'
MTB > name c4 'predict'
MTB > print c2 c3 c4 c1

ROW    values    probs        predict    observed

  1      0       0.342088        34          36
  2      1       0.374940        37          34
  3      2       0.197254        20          21
  4      3       0.066301         7           7
  5      4       0.015987         2           2
  6      5       0.002944         0
  7      6       0.000430         0
  8      7       0.000051         0
  9      8       0.000005         0
```

Fig 3.2 Calculation of the expected frequencies.

```
MTB > set c1
DATA> 5:25
DATA> end
MTB > pdf probabilities of c1 in c2;
SUBC> binomial n=25, p=0.042.
MTB > sum c2
     SUM    =    0.0034306
```

Fig 3.3 Calculation of $P(X \geq 5)$.

The expected value of a binomial random variable is np, and its variance is given by $np(1-p)$. Hence, the expected number of machines failing in a given day is estimated to be $np = 25 \times 0.042 = 1.05$ (as before).

As well as calculating probabilities, we can use a computer to *simulate* experiments. For example, we can generate random values from a binomial distribution with $n = 25$ and $p = 0.042$ and count how many are bigger than 4 (see Fig 3.4).

```
MTB > random 1825 values in c1;
SUBC> binomial n=25, p=0.042.
MTB > tally c1

        C1    COUNT
         0     618
         1     652
         2     372
         3     140
         4      35
         5       7
         6       1
        N=    1825
```

Fig 3.4 Simulation of a binomial distribution.

In this particular simulation, eight values exceeded 4.

Example I

An airline sells tickets for a plane with 30 seats. From experience the airline knows that there is a 20% chance that buyers of tickets will not turn up to claim their seats. On that basis the airline sells 35 tickets. What is the probability that more than 30 claim their seats? How many tickets do you think the airline should sell?

Are the requirements of Bernoulli trials satisfied in this case? There are two possible outcomes (arrival and non-arrival – we shall call 'arrival' a success). Independence may be a problem, as groups of people may travel together. Also the probabilities of different sorts of passengers may vary (e.g. business people, holiday-makers, etc.). We shall assume that the requirements are approximately satisfied. Here $n = 35$ and $p = 0.8$ (since an arrival is a success). The probability that more than 30 people arrive is calculated by summing all the probabilities of values 31 to 35 (see Fig 3.5).

So the probability that the flight is oversubscribed is 0.14349. What happens if the airline sells 34 or 33 tickets? Figure 3.6 shows the Minitab calculations.

```
MTB > set c1
DATA> 31:35
DATA> end
MTB > pdf probabilities of c1 in c2;
SUBC> binomial n=35, p=0.8.
MTB > print c2
C2
    0.0829678   0.0414839   0.0150850   0.0035494   0.0004056
MTB > sum c2
     SUM    =       0.14349
```

Fig 3.5 Summing the probabilities of values 31 to 35.

```
MTB > set c1
DATA> 31:34
DATA> end
MTB > pdf probabilities of c1 in c2;
SUBC> binomial n=34, p=0.8.
MTB > sum c2
    SUM    =    0.070006
MTB > set c1
DATA> 31:33
DATA> end
MTB > pdf probabilities of c1 in c2;
SUBC> binomial n=33, p=0.8.
MTB > sum c2
    SUM    =    0.026779
```

Fig 3.6 Calculations for $n = 34$ and $n = 33$.

For $n = 34$, the probability that too many passengers arrive is 0.070006 and for $n = 33$ it is 0.026779.

How many tickets do you think the airline should sell? Clearly the answer to this question will depend on the cost of compensation to disappointed passengers, the potential loss of custom due to dissatisfaction, the cost of running a flight that is not fully subscribed, etc. We may decide that it is acceptable to have disappointed passengers on up to 5% of flights, i.e. the probability that too many passengers arrive must not exceed 0.05. In that case, clearly $n = 34$ is too large, so 33 tickets should be sold.

Example 2

There are five copies of a course text in a university departmental library, and each copy can be booked out for one day at a time. If there are seven students on the course, and each (independently) has probability 0.6 of requiring a copy of the book on any given day, on how many days in a 12 week semester will at least one student find that no copies are available? Is it true that the mean number of books on loan per day is less than 4?

The number of students requesting the book may be assumed to be a binomial random variable, with $n = 7$ and $p = 0.6$. Figure 3.7 shows the Minitab calculations of the probabilities.

The probability that at least one student is disappointed is obtained by adding together the last two, $0.1306 + 0.028 = 0.1586$.

Call the number of books on loan N. What is the mean or expected value of N? Is it just the mean of the binomial distribution: $np = 4.2$? No, this is the mean of the *demand* (the mean number of students requesting the book), but not all of that demand is satisfied, so the mean of N must be less than this. You can find it by calculating the probability distribution of N. For the values 0 to 4 the probabilities are those of the binomial distribution, but $P(N = 5)$ is obtained by adding together

```
MTB > pdf;
SUBC> binomial n=7, p=0.6.
BINOMIAL WITH N =    7   P = 0.600000
         K              P( X = K)
         0                0.0016
         1                0.0172
         2                0.0774
         3                0.1935
         4                0.2903
         5                0.2613
         6                0.1306
         7                0.0280
```

Fig 3.7 Binomial probabilities with $n = 7$ and $p = 0.6$.

the last three of the "demand" probabilities: $0.2613 + 0.1306 + 0.028 = 0.4199$. So the pd of N is:

n	0	1	2	3	4	5
$P(N = n)$	0.0016	0.0172	0.0774	0.1935	0.2903	0.4199

Its expected value is therefore

$$(0 \times 0.0016) + (1 \times 0.0172) + (2 \times 0.0774) + (3 \times 0.1935) + (4 \times 0.2903)$$
$$+ (5 \times 0.4199) = 4.0132$$

So it is not true that the mean is less than 4.

EXERCISES ON 3.2

1. A manufacturer of electronic components sells 1 kΩ resistors in boxes of 10. A customer requires the components to meet a particular specification, and is happy to accept boxes containing no more than two which fail to comply. When the manufacturer tests 1000 resistors it finds that 95 do not meet the specification. If the manufacturer supplies 100 boxes to the customer, how many can it expect to be rejected?
2. In a multiple choice examination paper of 100 questions, in each question candidates are asked to select the one correct answer from four offered. They receive one mark for a correct answer and no marks for a wrong answer. If the pass mark is 35, what is the probability that a candidate will pass if he or she guesses all the answers (i.e. selects one of the four at random each time)?
3. A trainee glass-cutter has the job of cutting up large sheets of glass to make small panes of specific sizes. On average, for every 10 panes cut correctly the glass-cutter breaks one. From a sheet large enough to produce 12 panes, the glass-cutter needs to cut 10 correctly. What is the probability that the glass-cutter achieves this, assuming that the cuts are statistically independent?

48 Statistics

4. A batch of experimental electronic components is such that each component has (independently) only a 62% chance of functioning correctly. Two correctly functioning components are required for a space mission, but they can only be tested in orbit. How many of the components must be taken into space to be at least 99% certain of having two that function?

3.3 The Poisson distribution

Example 3

A local football team scores goals in such a way that any two time periods of equal length in the 90 minute match are equally likely to contain a goal, and are independent in the sense that the scoring of a goal in one period does not influence the likelihood of scoring a goal in another period. On average they score two goals per match. I want to work out the probabilities that they score zero, one, two, three or four goals in a match.

We shall call the number of goals scored in a match G. As a first attempt, we shall divide the 90 minutes into nine intervals of ten minutes each, and assume that each interval will contain zero or one goals. If we assume that the probability of scoring a goal in one interval is not affected by what happens in any other interval (the independence assumption) and that the probability is the same for each interval, G has a binomial distribution with $n = 9$. Noting that the average number of goals scored is 2, we deduce that $np = 2$, i.e. $p = 0.2222$. Figure 3.8 shows the binomial probabilities.

```
MTB > set c1
DATA> 0:4
DATA> end
MTB > pdf probabilities of c1 in c2;
SUBC> binomial n=9, p=0.2222.
MTB > print c2
C2
   0.104186    0.267874    0.306102    0.204042    0.087435
```

Fig 3.8 Binomial probabilities with $n = 9$ and $p = 0.2222$.

Our choice of nine intervals of 10 minutes each was arbitrary, and must be regarded as an approximation of the true situation since we have disregarded the possibility that more than one goal could be scored in one time interval. In an attempt to improve the approximation, we shall consider 18 five-minute time intervals. Now $n = 18$ and $p = 0.1111$. The corresponding probabilities are shown in Fig 3.9.

As may be expected, the probabilities have changed by a small amount. We shall try two more approximations. First 90 one-minute intervals; see Fig 3.10.

Again, the probabilities have changed by a small amount. Finally 5400 one-second intervals; see Fig 3.11.

```
MTB > pdf probabilities of c1 in c2;
SUBC> binomial n=18, p=0.1111.
MTB > print c2
C2
   0.120047   0.270076   0.286923   0.191261   0.089643
```

Fig 3.9 Binomial probabilities with $n = 18$ and $p = 0.1111$.

```
MTB > pdf probabilities of c1 in c2;
SUBC> binomial n=90, p=0.02222.
MTB > print c2
C2
   0.132344   0.270675   0.273723   0.182464   0.090186
```

Fig 3.10 Binomial probabilities with $n = 90$ and $p = 0.02222$.

```
MTB > pdf probabilities of c1 in c2;
SUBC> binomial n=5400, p=0.00037037.
MTB > print c2
C2
   0.135285   0.270671   0.270721   0.180480   0.090223
```

Fig 3.11 Binomial probabilities with $n = 5400$ and $p = 0.00037037$.

The changes in the resulting probabilities are now very small. At this point, you may reasonably object that the independence assumption is no longer valid as it is physically impossible to score two goals in adjacent seconds!

Mathematically, what we have done is to consider the effect on a particular set of binomial probabilities when n is allowed to get arbitrarily large while p gets arbitrarily small ($p > 0$), with the constraint that we keep the quantity np constant (in this case $np = 2$).

The result is a new distribution called the **Poisson distribution**, named after the French mathematician Simeon-Denis Poisson (1781–1840). The relevant probabilities for our football example are calculated in Fig 3.12.

```
MTB > pdf probabilities of c1 in c2;
SUBC> poisson 2.
MTB > print c2
C2
   0.135335   0.270671   0.270671   0.180447   0.090224
```

Fig 3.12 Poisson probabilities with mean 2.

Compare these with the binomial probabilities with $n = 5400$, $p = 0.00037037$. Note that the only parameter you need to specify for the Poisson distribution is its mean, which we shall denote by the Greek letter μ.

We can calculate the probabilities associated with all possible outcomes of a Poisson distribution, provided we know μ, using the formula

$$P(X = x) = \frac{e^{-\mu}\mu^x}{x!} \quad (x = 0, 1, 2, \ldots)$$

These probabilities can also be found in books of tables, or calculated by a computer package.

Example 4

Dataset 4 (Appendix A) contains the times of all goals scored in the English Football League on a particular Saturday. Do these data appear to confirm our theories about the Poisson distribution? We might expect the number of goals scored each minute to be a random variable with a Poisson distribution.

To test this theory, we can first summarise the data in a frequency table:

Goals	0	1	2	3	4
Number of minutes	39	31	13	5	2

In other words, of the 90 minutes played, there were 39 in which no goals were scored, 31 in which one goal was scored, etc. Are these the sort of frequencies we would expect from a Poisson distribution? Before we can make a meaningful comparison, we need to know the *mean* of the Poisson distribution. This should be the mean number of goals scored per minute and can be estimated from the data. In all, 80 goals were scored in 90 minutes, so the mean is $80/90 = 0.8889$.

The Poisson distribution with mean 0.8889 gives the following probabilities:

Value	0	1	2	3	4	5
Probability	0.4111	0.3654	0.1624	0.0481	0.0107	0.0019

So, for example, the probability that two goals are scored in a particular minute is 0.1624. In 90 minutes (90 'trials') we would therefore expect *on average* $90 \times 0.1624 = 14.616$ minutes when two goals are scored. (This is, in fact, another application of the Binomial distribution i.e. $n = 90$ trials with constant probability of success $p = 0.1624$, having mean $np = 14.616$.)

Applying this procedure to each of the Poisson probabilities, we can draw up a table of *expected* numbers of minutes for each number of goals:

Goals	0	1	2	3	4	5
Observed minutes	39	31	13	5	2	0
Expected minutes	37.0	32.9	14.6	4.3	1.0	0.2

The observed values are included for comparison. For this particular set of real data they match very well, lending support to our theory that the number of goals scored per minute is a Poisson random variable.

We can use the Poisson distribution to model events that occur randomly in time or space, e.g. the number of telephone calls handled by a switchboard every five minutes, the number of flaws per metre in a bolt of material. The requirements are:

- The number of events in any interval must be independent of any other interval.
- The average number of events is proportional to the size of the interval.
- Events cannot occur simultaneously (in time or space).

Example 5

A shift supervisor records the number of accidents each week for machine operators in a workshop. Draw up a frequency table of the data and calculate the sample mean. Decide whether the Poisson distribution provides a suitable model for these data. If so, predict the number of weeks that will result in 10 or more accidents in the next 10 years (520 weeks).

Data: 2, 0, 4, 1, 4, 1, 4, 3, 6, 1, 4, 10, 6, 6, 5, 3, 3, 4, 1, 3, 2, 6, 2, 5, 1, 6, 4, 7, 2, 2

The Minitab calculations are shown in Fig 3.13. We must compare these frequencies with the observed ones in the tally above. They do not match well, but there are no large discrepancies. In Fig 3.14 we calculate the probability of 10 or more accidents by calculating the probability of zero to nine and subtracting the result from 1.

In 520 weeks, we would expect $520 \times 0.00402433 = 2.09$ weeks in which 10 or more accidents occur (i.e. it will happen roughly twice in 10 years).

Example 6

A car hire firm has four cars for hire on a daily basis. Suppose that the daily demand for car hire has a Poisson distribution with mean 3.5 cars per day. If the manager recorded the number of cars hired each day for a large number of days, what would you expect the mean of the resulting sample to be? In 100 days, on how many occasions would you expect all four cars to be hired out?

The probabilities associated with hiring zero, one, two or three cars are the Poisson probabilities of the demand. The probability that four cars are hired is equal to the probability that the demand is four *or more*. This is obtained by summing the probabilities of 0, 1, 2 and 3, and subtracting the result from 1. Then the mean of the resulting distribution is obtained by multiplying the values (0–4) by their corresponding probabilities and summing. The Minitab calculations are shown in Fig 3.15.

The mean number of cars hired per day is 2.9764. The probability that four cars are hired is 0.463367. So in 100 days we would expect about 46 days on which all four cars are hired out.

Poisson approximation to the binomial distribution

As can be seen from Example 3, the probabilities calculated from the Poisson distribution are very similar to those calculated from a binomial distribution with

```
MTB > set c1
DATA> 2 0 4 1 4 1 4 3 6 1
DATA> 4 10 6 6 5 3 3 4 1 3
DATA> 2 6 2 5 1 6 4 7 2 2
DATA> end
MTB > tally c1

         C1    COUNT
         0      1
         1      5
         2      5
         3      4
         4      6
         5      2
         6      5
         7      1
        10      1
        N=     30

MTB > mean c1 in k1
    MEAN    =       3.6000

MTB > set c2
DATA> 0:10
DATA> end
MTB > pdf probabilities of c2 in c3;
SUBC> poisson k1.

MTB > let c4=round(c3*30)
MTB > print c4

 1    3    5    6    6    4    2    1    1    0    0
```

Fig 3.13 Minitab calculations for Example 5.

```
MTB > set c2
DATA> 0:9
DATA> end
MTB > pdf probabilities of c2 in c3;
SUBC> poisson k1.
MTB > let k2=1-sum(c3)
MTB > print k2
K2        0.00402433
```

Fig 3.14 The probability of 10 or more accidents.

```
MTB > set c1
DATA> 0:3
DATA> end
MTB > pdf probabilities of c1 in c2;
SUBC> poisson 3.5.
MTB > let c2(5)=1-sum(c2)
MTB > let c1(5)=4
MTB > let c3=c1*c2
MTB > sum c3
    SUM    =    2.9764
MTB > print c2
C2
   0.030197    0.105691    0.184959    0.215785    0.463367
```

Fig 3.15 Minitab calculations for Example 6.

large n and small p, provided the Poisson parameter μ is set equal to np. The Poisson distribution may therefore be used as an approximation to the binomial distribution, if n is large and p is small.

EXERCISES ON 3.3

1. The number of oil tankers arriving at a certain refinery each day constitutes a Poisson process with parameter two tankers per day. Present port facilities can service three tankers a day. If more than three tankers arrive in a day then the tankers in excess of three must be sent to another port.
   (a) What is the expected number of tankers arriving per day?
   (b) What is the most probable number of tankers arriving daily?
   (c) What is the expected number of tankers serviced daily?
   (d) What is the expected number of tankers turned away daily?
   (e) If port facilities must cope with daily demand with probability at least 0.99 then to what extent must they be extended?
2. A typist types pages of text with 65 characters per line and 55 lines per page. Suppose $p = 0.005$ is the probability that any particular keystroke is incorrect. Let X be the total number of errors on a page. Assuming that the keystrokes are stochastically independent, what is $P(X < 4)$? (The number of errors will follow a binomial distribution with very large n and small p. Use the Poisson distribution as an approximation.)
   Now suppose that a more skilful typist takes over the job, so that $p = 0.001$ (on average, one mistake per 1000 keystrokes). With this level of skill, what percentage of pages will have three errors or fewer?
3. The college library has five copies of the text book *How to do Stats*, which is so popular that copies can only be borrowed for an hour at a time. The demand for this text has a Poisson distribution, with a mean value of four students per hour.
   (a) What is the probability that during a particular hour all five copies will be on loan?

(b) What is the expected number of copies on loan in any hour?
(c) In a 72 hour working week, the number of copies in use each hour gave the following frequency table:

Copies	0	1	2	3	4	5
Hours	0	3	11	12	15	31

Are these values consistent with the assumption that demand is Poisson with mean 4?

3.4 The hypergeometric distribution

A youth club leader entering a team for a competition claims to give equal opportunities to boys and girls. From a club of 50 members, 22 of whom are girls, the leader chooses a team of eight boys and two girls. Are the girls justified in feeling aggrieved?

You can tackle this problem by asking the question 'how likely is it that as few as two girls would be chosen if the team were selected at random?' If this probability turns out to be small, then the girls' indignation may well be justified.

At first sight this looks like another binomial example. We can define the selection of a girl as a 'success', and there are 10 selections or 'trials'. But the probability of a success is not the same for each trial, as the population from which the leader is choosing is finite and the individuals selected are not being replaced. Furthermore, the composition of the remaining population after each choice will depend upon the choices already made, so the choices are not independent. For a very large club membership we could ignore these problems and use the binomial distribution as a good approximation, but for this situation we need a new distribution.

The number of 'successes' in a sample of size n drawn from a population of size N containing k successes has a **hypergeometric** distribution with parameters N, n and k.

If X is a random variable with a hypergeometric distribution, X has probability density given by the formula

$$P(X = x) = \binom{N-k}{n-x}\binom{k}{x} \bigg/ \binom{N}{n}$$

X will have mean nk/N and variance

$$\frac{n(k/N)(1 - k/N)(N - n)}{(N - 1)}$$

For the youth club problem, let the number of girls chosen be G.

$$P(G=2) = \binom{28}{8}\binom{22}{2} \bigg/ \binom{50}{10} = 0.0699$$

$$P(G=1) = \binom{28}{9}\binom{22}{1} \bigg/ \binom{50}{10} = 0.0148$$

$$P(G=0) = \binom{28}{10}\binom{22}{0} \bigg/ \binom{50}{10} = 0.0013$$

So the probability that as few as two girls are chosen (i.e. zero, one or two) is obtained by adding up these three values: $0.0699 + 0.0148 + 0.0013 = 0.086$.

This is less than one chance in 10, so there is some evidence that the boys have been favoured.

Binomial approximation to the hypergeometric distribution

If n is small compared with N (say $n/N < 0.05$) we can get good results by using the binomial distribution with n trials and $p = k/N$.

Example 7

From a town of 1000 adults, a jury of 12 is selected at random. What is the probability that one or more of the 15 members of the town rugby team is selected for jury service?

Here $N = 1000$ and $n = 12$ so we shall use the binomial approximation with $n = 12$ and $p = 15/1000 = 0.015$.

$$p_0 = P \text{ (no rugby player selected)} = (1-p)^n = 0.985^{12} = 0.8341.$$

So P (1 or more selected) $= 1 - 0.8341 = 0.1659$.

Using the hypergeometric distribution,

$$p_0 = \binom{1000-15}{12}\binom{15}{0} \bigg/ \binom{1000}{12} = 0.8333$$

As you can see, the approximation is very good.

EXERCISE ON 3.4

1. In a state lottery, you choose six numbers from 60 possible values and you win a prize if you get three or more of the six numbers drawn at random. What is the probability that you win a prize? What is the probability that you win/share the jackpot by getting all six numbers?

3.5 The geometric distribution

Returning to the laboratory of 25 PCs of Section 3.2, we note from the Minitab output that the probability that there are no failures in a given day is 0.3421. What is the probability that the *first* failure-free day is day 5? Put another way, let D be the number of the first failure-free day. Then D is a random variable, and our original question involves calculating $P(D = 5)$. This is an example of the **geometric distribution**. D has a geometric distribution if D represents the number of Bernoulli trials up to and including the one in which the first success occurs. The only parameter required is p, the probability of success on each trial. D takes the values $1, 2, 3, \ldots$ (any positive integer) and has the probability density

$$P(D = d) = p(1 - p)^{d-1}$$

D has mean $1/p$ and variance $(1 - p)/p^2$.

For $p = 0.3421$, D has the probability distribution

d	1	2	3	4	5	6	7	8
$P(D=d)$	0.3421	0.2251	0.1481	0.0974	0.0641	0.0422	0.0277	0.0183

So the probability that the first failure-free day is day 5 equals 0.0641.

How many days will pass *on average* before the first failure-free day?

The mean value of D is $1/p = 2.9231$. So the average number of days before the first failure-free day is 1.9231.

EXERCISE ON 3.5

1. A toy manufacturer introduced a new range of model kits aimed at customers who were currently buying a similar range of ready assembled models. After release, the progress of sales of the new item was monitored. Data collected at one of the major retail outlets gave the following summary. This gives the frequency distribution of the values n of N, the number of goods sold up to and including the sale of the next single item from the new range.

n	1	2	3	4	5	6	7	8	9	10
Frequency	126	126	60	28	25	8	11	1	5	1

Assuming that the actual item sold in any single sale is independent of any other items sold and that the probability of a new item being sold is p, a constant, a probability model for N is

$$P(N = n) = p(1 - p)^{n-1} \quad (n = 1, 2, \ldots)$$

An estimate of p is $p = 391/983$. What does p represent, i.e. how has it been constructed from the data? Using the estimate, calculate expected frequencies corresponding to the observed frequencies above. Comment on the results.

3.6 The negative binomial distribution

An obvious generalisation of the question which led us to the geometric distribution is 'What is the probability that the rth failure-free day is day d?' (where r may be any positive integer).

Let D_r be the number of the rth failure-free day. Then D_r is an example of a random variable with a **negative binomial distribution**. D_r has a negative binomial distribution if D_r represents the number of Bernoulli trials up to and including the one in which the rth success occurs. The parameters are r and p (the probability of success on each trial). D_r takes the values $r, r+1, r+2, \ldots$ and has the probability density

$$P(D_r = d) = \binom{d-1}{r-1} p^r (1-p)^{d-r}$$

D_r has mean r/p and variance $r(1-p)/p^2$. When $r = 1$ this is, of course, the geometric distribution.

What is the probability that there will be at least three failure-free days in the first seven days? This is equivalent to asking for the probability that D_3 is less than or equal to 7. The probability distribution of D_3 when $p = 0.3421$ is

d	3	4	5	6	7	8	9	10
$P(D_3 = d)$	0.0400	0.0790	0.1040	0.1140	0.1125	0.1036	0.0909	0.0769

Summing the first five of these we get $P(D_3 = 7) = 0.4495$.

EXERCISE ON 3.6

1. If a fair coin is tossed until five heads are obtained, what is the probability that this occurs on the 10th toss? What is the probability that this occurs *before* the 10th toss?

Summary

For discrete random variable X, taking values x_i with probabilities p_i,

$$E(X) = \sum_{i=1}^{n} x_i p_i \qquad Var(X) = E(X^2) - [E(X)]^2$$

Binomial distribution

X has a *binomial distribution* with parameters n and p if X is the number of successes in n independent trials with constant probability of success p.

$$P(X = x_i) = \binom{n}{x_i} p^{x_i} (1-p)^{n-x_i} \quad (x_i = 0, 1, 2, \ldots, n)$$

$$E(X) = np \qquad Var(X) = np(1-p)$$

Poisson distribution

If events occur randomly in time or space, the number of events in an interval will have a *Poisson distribution* with parameter μ if

- The number of events in any interval is independent of any other interval.
- The average number of events is proportional to the size of the interval.
- Events cannot occur simultaneously (in time or space).

$$P(X = x_i) = \frac{e^{-\mu}\mu^{x_i}}{x_i!} \quad (x_i = 0, 1, 2, \ldots)$$

$$E(X) = \mu \quad Var(X) = \mu$$

A Poisson distribution with parameter μ may be used to approximate a binomial distribution with parameters n and p provided n is large, p is small and $\mu = np$.

Hypergeometric distribution

Discrete random variable X has a *hypergeometric distribution* with parameters N, n and k if X is the number of successes in a sample of size n drawn from a population of size N containing k successes.

$$P(X = x) = \binom{N-k}{n-x}\binom{k}{x} \Big/ \binom{N}{n}$$

$$E(X) = nk/N$$

$$Var(X) = \frac{n(k/N)(1 - k/N)(N - n)}{(N - 1)}$$

A hypergeometric distribution with parameters N, n and k may be approximated by a binomial distribution with parameters n and p provided n is small compared with N and $p = k/N$.

Geometric distribution

Discrete random variable X has a *geometric distribution* if X is the number of Bernoulli trials, with probability of success p, up to and including the one in which the first success occurs.

$$P(X = x) = p(1 - p)^{x-1} \quad (x = 1, 2, 3, \ldots)$$

$$E(X) = \frac{1}{p} \quad Var(X) = \frac{1-p}{p^2}$$

Negative binomial distribution

Discrete random variable X has a *negative binomial distribution* with parameters r and p if X is the number of Bernoulli trials up to and including the one in which the rth success occurs.

$$P(X = x) = \binom{x-1}{r-1} p^r (1 - p)^{x-r} \quad (x = r+1, r+2, r+3, \ldots)$$

$$E(X) = \frac{r}{p} \quad Var(X) = \frac{r(1-p)}{p^2}$$

3.7 Continuous distributions

Suppose the number of customers arriving at a supermarket checkout each minute has a Poisson distribution with mean 2. What will be the average value of T, the *time between arrivals*?

Clearly if we have an average of two customers per minute, the time between their arrivals must average out at 30 seconds. But what sort of distribution does T

follow? The first thing we notice is that T can take *any* positive value – not just whole numbers. The set of possible values is not only infinite, it is *uncountable*. This sort of random variable is called a **continuous** random variable, and we describe its probability structure by means of a **probability density function** (pdf) or a **cumulative distribution function** (cdf).

The probability density function

The pdf is a function with the following property: the probability that the value of the random variable lies in a given interval is equal to the *area* under the pdf corresponding to that interval.

The pdf for our random variable T, the time between arrivals, is

$$f(t) = 2e^{-2t} \quad (t > 0)$$
$$= 0 \quad \text{(otherwise)}$$

The graph of this function is shown in Fig 3.16.

What is the probability that the time between arrivals, T, is between 30 seconds and a minute? We can answer this by finding the area under the curve between 0.5 and 1.0 (the area shaded in Fig 3.16). One way to do this is by drawing an accurate graph, by hand or using computer graphics facilities, and estimating the area by counting squares. Another way is by *integration*.

The integral of the pdf between 0.5 and 1 gives the area under the curve and hence the required probability:

$$P(0.5 < T < 1) = \int_{0.5}^{1} 2e^{-2t} dt = [-e^{-2t}]_{0.5}^{1} = e^{-1} - e^{-2} = 0.2325$$

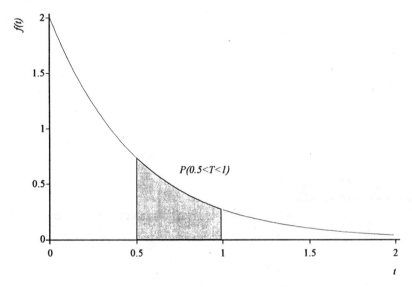

Fig 3.16 The pdf of $f(t) = 2e^{-2t}$.

The cumulative distribution function

A simpler approach involves performing the integration once only to find $P(T \leq t)$ for any t. This is called the **cumulative distribution function** (cdf).

$$F(t) = P(T \leq t) = \int_0^t 2e^{-2x}dx = [-e^{-2x}]_0^t = 1 - e^{-2t} \quad (t > 0)$$

$$= 0 \quad \text{(otherwise)}$$

This function is all we need to solve the problem for any interval, since, for example,

$$P(0.5 < T \leq 1) = P(T \leq 1) - P(T \leq 0.5) = F(1) - F(0.5)$$
$$= (1 - e^{-2}) - (1 - e^{-1}) = e^{-1} - e^{-2} = 0.2325$$

Clearly, if we know the form of the cdf of a distribution, we can calculate any probabilities without graphs or integration.

3.8 The mean and variance of a continuous distribution

We calculate the mean of a continuous distribution from its probability density function:

$$E(X) = \int_{-\infty}^{\infty} xf(x)dx$$

This is analogous to the discrete procedure of multiplying the possible values by their corresponding probabilities and summing.

For our random variable T, note that $f(t)$ is zero for $t < 0$, so

$$E(T) = \int_0^{\infty} t \cdot 2e^{-2t}dt = [-te^{-2t} - \tfrac{1}{2}e^{-2t}]_0^{\infty} = \tfrac{1}{2}$$

We calculate the variance of a continuous distribution from

$$Var(X) = E(X^2) - [E(X)]^2 \quad \text{and} \quad E(X^2) = \int_{-\infty}^{\infty} x^2 f(x) dx$$

For T,

$$E(T^2) = \int_0^{\infty} t^2 \cdot 2e^{-2t} dt = [-e^{-2t}(t^2 + t + \tfrac{1}{2})]_0^{\infty} = \tfrac{1}{2}$$

$$Var(T) = E(T^2) - [E(T)]^2 = \tfrac{1}{2} - \tfrac{1}{4} = \tfrac{1}{4}$$

EXERCISES ON 3.8

1. Random variable X has the probability density function defined by

$$f(x) = ax \quad (0 \leq x \leq 1)$$
$$= \frac{a(3-x)}{2} \quad (1 \leq x \leq 3)$$
$$= 0 \quad \text{otherwise}$$

Calculate numerical values for:
(a) the value of a;
(b) $E(X)$;
(c) $Var(X)$.
2. Random variable X is chosen at random from the interval (a, b). This is called a **uniform distribution**. Write down a formula for its pdf, and sketch it. Calculate the mean and variance of X.

3.9 The exponential distribution

If a random variable X, representing the number of arrivals in a unit time period, has a Poisson distribution with mean μ, then the time between arrivals, T, will have an **exponential distribution** with mean μ^{-1}. The pdf of this distribution is

$$f(t) = \mu e^{-\mu t}$$

and its cumulative distribution function is

$$F(t) = P(T \leq t) = 1 - e^{-\mu t}$$

It has mean $E(T) = \mu^{-1}$ and variance $Var(T) = \mu^{-2}$.

Note that, while the Poisson distribution has its mean and *variance* equal, the exponential distribution has its mean and *standard deviation* equal.

The distribution described in Section 3.7 is exponential with $\mu = 2$.

Example 8

Returning to Dataset 4, the times of English Football League goals, do the times *between* goals appear to follow an exponential distribution?

To test this theory, we first calculate the differences between successive goal times (taking the first goal time as the first 'difference'). The ordered goal times are:

1.1 1.8 2.4 2.8 3.8 4.7 6.4 6.5 7.1 7.7 9.6 9.7 11.0 13.1 16.3 23.3 23.8 24.4 24.6 26.3 27.0 27.1 27.3 31.0 34.1 36.0 37.9 41.1 42.1 42.2 44.1 46.6 47.0 47.8 48.6 49.0 49.1 49.4 51.9 53.5 54.6 55.7 56.0 56.9 57.1 57.1 57.5 57.9 58.4 58.8 59.6 59.7 61.7 62.8 63.5 67.3 67.7 70.0 70.8 71.1 71.5 73.3 74.8 75.5 75.7 80.9 81.0 81.5 84.0 84.5 85.2 86.3 87.0 87.4 87.6 87.6 87.6 88.8 89.3 89.4

Taking consecutive differences:

0.7 0.6 0.4 1.0 0.9 1.7 0.1 0.6 0.6 1.9 0.1 1.3 2.1 3.2 7.0 0.5 0.6 0.2 1.7 0.7 0.1 0.2 3.7 3.1 1.9 1.9 3.2 1.0 0.1 1.9 2.5 0.4 0.8 0.8 0.4 0.1 0.3 2.5 1.6 1.1 1.1 0.3 0.9 0.2 0.0 0.4 0.4 0.5 0.4 0.8 0.1 2.0 1.1 0.7 3.8 0.4 2.3 0.8 0.3 0.4 1.8 1.5 0.7 0.2 5.2 0.1 0.5 2.5 0.5 0.7 1.1 0.7 0.4 0.2 0.0 0.0 1.2 0.5 0.1

From this data, we produce the following frequencies for 0–0.4, 0.5–0.9, etc.:

28 21 9 9 3 3 3 2 0 0 1 0 0 0 1

We can calculate the probabilities for these categories from the cdf of the exponential distribution, but first we need to decide exactly what range of times would have been allocated to each category. The times are recorded to the nearest tenth of a minute. It is not clear whether the raw observations were truncated or rounded, but we shall assume rounding, i.e. the category 0–0.4 contains values between 0 and 0.45, and 0.5–0.9 contains values between 0.45 and 0.95 and so on.

Now, 80 goals in 90 minutes gives an average time difference of $90/80 = 1.125$, i.e. $\mu = 0.8889$. The first probabilities are therefore

$$P(0 < T < 0.45) = F(0.45) - F(0) = (1 - e^{-0.45\mu}) - (1 - e^0) = 0.3297$$
$$P(0.45 < T < 0.95) = F(0.95) - F(0.45) = (1 - e^{-0.95\mu}) - (1 - e^{-0.45\mu})$$
$$= 0.2405$$

Figure 3.17 shows the calculation of these probabilities in Minitab, followed by a prediction of the number in each category, obtained by multiplying the probabilities by 80.

Comparing these calculated frequencies with the observed frequencies we see that the fit is good. The only major discrepancy is the seven minute gap in the goal times between 16.3 and 23.3 minutes, which produces the one outlying value in our frequency table.

```
MTB > set c1
DATA> 0:16
DATA> end
MTB > let c2=(c1*.5)-.05
MTB > let c2(1)=0
MTB > print c2

C2
   0.00    0.45    0.95    1.45    1.95    2.45    2.95    3.45    3.95
   4.45    4.95    5.45    5.95    6.45    6.95    7.45    7.95

MTB > cdf c2 in c3;
SUBC> exponential 1.125.
MTB > print c3

C3
   0.000000   0.329680   0.570204   0.724423   0.823306   0.886707
   0.927359   0.953424   0.970136   0.980852   0.987723   0.992128
   0.994953   0.996764   0.997925   0.998670   0.999147

MTB > differences c3 in c4
MTB > print c4

C4
          *    0.329680   0.240524   0.154219   0.098882   0.063401
   0.040652   0.026065   0.016712   0.010716   0.006871   0.004405
   0.002825   0.001811   0.001161   0.000745   0.000477

MTB > let c5=round(c4*80)
MTB > print c5

C5
        *     26     19     12      8      5      3      2      1      1     1
        0      0      0      0      0      0
```

Fig 3.17 Minitab calculation of probabilities and predicted numbers.

EXERCISE ON 3.9

1. Data on the failure times of certain electronic components gave rise to the following frequency table. Assuming an exponential failure time distribution with probability density function

$$\frac{1}{\mu} e^{-t/\mu} \quad (t > 0)$$

where μ = expected failure time, calculate the expected frequencies and hence give an assessment of the appropriateness of the model.

Class interval	0–500	500–1000	1000–1500	1500–2000	2000–3000
Frequency	324	109	45	15	7

Hint: Start by calculating the mean failure time and equate it to μ.

3.10 The normal distribution

If we collect data on human, animal or plant characteristics such as height, weight, life expectancy, IQ, etc., and plot histograms, the result is very often an approximately bell-shaped distribution called the **normal distribution**. Of course, the different characteristics will have different means and different variances or standard deviations, but the overall shape of the distribution is remarkably consistent. We observe similar results when we make repeated observations of the same quantity, but with experimental errors – the errors also follow the normal distribution.

> The discovery of the bell-shaped normal curve, also known as the Gaussian distribution, is usually attributed to Karl Gauss (1777–1855), regarded by many as one of the greatest mathematicians of all time. It seems likely, however, that the earliest work in this area was done by Abraham de Moivre (1667–1754), a Frenchman who spent most of his adult life in Britain and was made a Fellow of the Royal Society in 1697.

The standard normal distribution

We call the particular normal distribution, for which the mean is zero and the standard deviation (or variance) is one, the **standard normal distribution**. If Z is a random variable with a standard normal distribution, we indicate this by writing $Z \sim N(0, 1)$. The pdf of this distribution is

$$\phi(z) = \frac{1}{\sqrt{(2\pi)}} \exp(-\tfrac{1}{2}z^2) \quad (-\infty < z < \infty)$$

Figure 3.18 is a graph of $\phi(z)$, showing the characteristic bell shape.

This pdf is difficult to integrate, so we usually obtain the cdf, $\Phi(z)$, from statistical computer packages or from normal distribution tables.

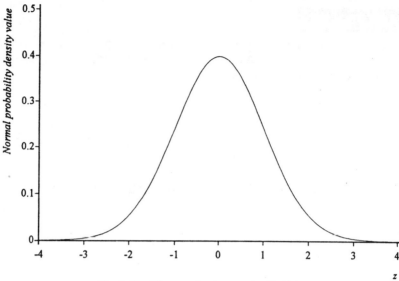

Fig 3.18 The standard normal pdf, $\phi(z)$.

Example 9

If $Z \sim N(0, 1)$, what percentage of observations of Z would you expect to lie in the interval $(-2, +2)$?

We know that $P(Z \leq 2) = \Phi(2)$ and $P(Z \leq -2) = \Phi(-2)$. So $P(-2 < Z \leq 2) = \Phi(2) - \Phi(-2)$. (As this is a continuous distribution we can ignore the difference between $P(Z \leq 2)$ and $P(Z < 2)$.)

We can now evaluate this expression using statistical tables, or in Minitab (see Fig 3.19). (Note that Minitab assumes $N(0, 1)$ if no distribution is specified.)

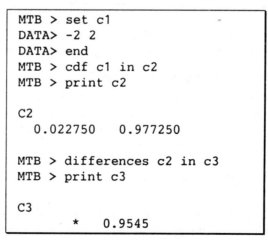

Fig 3.19 Evaluation of a normal probability using Minitab.

We can describe the result of Example 9 by saying '95.45% of the distribution lies within two standard deviations of the mean'. An important property of the normal distribution is that this statement is true for *all* normal distributions, not just $N(0, 1)$.

The general normal distribution

If X is a normally distributed random variable with mean μ and variance σ^2 we write $X \sim N(\mu, \sigma^2)$. The pdf of X is then

$$f(x) = \frac{1}{\sqrt{2\pi\sigma^2}} \exp\left[-\frac{1}{2}\left(\frac{x-\mu}{\sigma}\right)^2\right] \quad (-\infty < x < \infty)$$

To evaluate the cdf of X at a point x_0, we must integrate:

$$F(x_0) = \int_{-\infty}^{\frac{x_0-\mu}{\sigma}} \frac{1}{\sqrt{2\pi}} \exp(-\frac{1}{2}z^2)dz = \Phi\left(\frac{x_0-\mu}{\sigma}\right)$$

So we can calculate the cdf of any normally distributed random variable from the cdf of the $N(0, 1)$ distribution (provided that we know the mean and variance).

Example 10

Response times for a computer database program are normally distributed with mean 46.2 seconds and standard deviation 13.7 seconds. Users become irritated and complain to the database manager if the response time exceeds a minute. If 1000 customers access the database, how many complaints may be expected?

Call the response time T; then $T \sim N(46.2, 13.7^2)$. The probability that T exceeds a minute is given by

$$P(T > 60) = 1 - P(T < 60) = 1 - \Phi\left(\frac{60-46.2}{13.7}\right) = 1 - \Phi(1.0073)$$

In this calculation, we have found the cdf of T, $P(T < 60)$, by transforming to an $N(0, 1)$ random variable (subtracting the mean, 46.2, and dividing by the standard deviation, 13.7). The value of $\Phi(1.0073)$ can now be found using statistical tables or a computer package, e.g. the Minitab output of Fig 3.20.

```
MTB > let k1=(60-46.2)/13.7
MTB > cdf k1
     1.0073    0.8431
```

Fig 3.20 Minitab calculation of $\Phi(1.0073)$.

So $P(T > 60) = 1 - 0.8431 = 0.1569$.

In fact, most computer packages do the whole calculation for you, if you supply the necessary parameter values, as in Fig 3.21.

```
MTB > cdf 60;
SUBC> normal 46.2 13.7.
    60.0000      0.8431
```

Fig 3.21 Minitab calculation of $P(T < 60)$.

Note that Minitab requires you to supply the mean and the standard deviation. Some packages require the mean and the variance – you must check which combination is needed for your package.

The probability is also the *proportion* of T values expected to exceed a minute. So out of 1000 customers the manager may expect 1000×0.1569 or about 157 complaints.

EXERCISES ON 3.10

1. The width of a slot of a duralumin forging is (in inches) normally distributed with mean 0.9000 and standard deviation 0.0030. The specification limits were given as 0.9000 plus or minus 0.005. What percentage of forgings will be defective? What is the maximum allowed value of the standard deviation that will permit no more than 1 in 1000 defectives when the width is normally distributed with mean 0.9000?

2. An automatic milk dispenser is used to fill pint milk bottles. The amount of milk dispensed is normally distributed with mean 20.5 fluid ounces and standard deviation 0.2 fluid ounces. What is the probability that a randomly selected bottle contains less than a pint (20 fluid ounces) of milk?

3. The following are observations of a random variable X which is thought to have a normal distribution:

4.74	3.42	1.95	2.06	3.28	1.44	2.94	2.88	2.60	3.09
4.78	2.11	2.99	4.16	2.61	2.79	3.51	1.42	3.58	0.36
1.33	2.24	4.05	3.33	−1.57	1.85	0.83	1.10	1.53	3.00
1.18	3.36	4.30	2.88	2.00	−0.77	4.01	2.17	4.31	2.67
3.49	2.62	1.58	4.75	−1.92	4.41	4.49	3.69	0.72	1.79

Calculate the mean and variance, and use your results (assuming a normal distribution) to estimate $P(X > 5)$. Is this probability consistent with the data?

The normal approximation to the binomial distribution

A binomially distributed random variable with parameters n and p has properties very similar to a normally distributed random variable with mean np and variance $np(1 - p)$ provided n is large and $\hat{p}$ is not too close to zero or one. A good 'rule of thumb' is: use the normal approximation if np and $n(1 - p)$ are both at least five.

As the binomial distribution is discrete and the normal distribution is continuous, we will get a better approximation if we make a 'continuity correction'. Figure 3.22 shows the pdf of the distribution $N(10, 5)$ superimposed on a histogram of the binomial distribution with $n = 20$ and $p = 0.5$.

The column of the histogram representing the probability of 14 is approximated by the area under the curve between 13.5 and 14.5. Figure 3.23 shows the

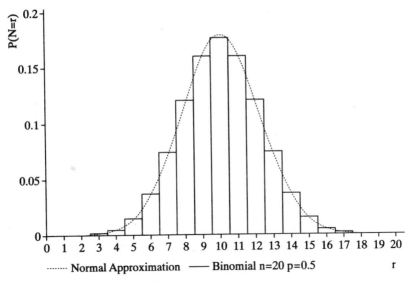

Fig 3.22 Bi(20,0.5) and N(10,5).

```
MTB > pdf 14;
SUBC> binomial n=20, p=0.5.
   K        P( X = K)
  14.00        0.0370

MTB > set c1
DATA> 13.5 14.5
DATA> end
MTB > let k1=sqrt(5)
MTB > cdf c1 in c2;
SUBC> normal 10 k1.
MTB > differences c2 in c3
MTB > print c3

C3
       *   0.0366768
```

Fig 3.23 The normal approximation to $P(X = 14)$ for $X \sim Bi(20, 0.5)$.

calculation of this probability, using the binomial distribution (for the accurate value) and using the normal approximation.

So the binomial probability is 0.0370, and the normal approximation is 0.0367, which matches to three decimal places.

How do we calculate $P(7 \leq X < 12)$ for $X \sim Bi(20, 0.5)$? The lower limit includes equality, so the continuous approximation starts at 6.5. The upper limit is really equality with 11 (as 12 is excluded by the strict inequality) so the continuous approximation ends at 11.5. Using the normal approximation:

$$P(7 \leq X < 12) = P(6.5 < Z < 11.5) \quad \text{for} \quad Z \sim N(10, 5).$$

The Minitab calculations are shown in Fig 3.24.

```
MTB > set c1
DATA> 7:11
DATA> end
MTB > pdf probabilities of c1 in c2;
SUBC> binomial n=20, p=0.5.
MTB > sum c2
   SUM   =   0.69062

MTB > let k1=sqrt(5)
MTB > set c3
DATA> 6.5 11.5
DATA> end
MTB > cdf c3 in c4;
SUBC> normal 10 k1.
MTB > differences c4 in c5
MTB > print c5

C5
    *    0.69007
```

Fig 3.24 The normal approximation to $P(7 \leq X < 12)$ for $X \sim Bi(20, 0.5)$.

The binomial probability is 0.6906, and the normal approximation is 0.6901, again matching to three decimal places.

Example 11

In an examination there are 100 questions, to which the answer is always either 'true' or 'false'. After the examination a girl student claims that she knew the answers to only 10 questions and guessed the rest. When the examination is marked, her score is 70 correct answers. Do you think she really guessed as many as she claimed?

If her claim is correct, she guessed 90 answers and 60 of them were correct. The probability of a guessed answer being correct is 0.5, so the number of correct answers N should be binomial with $n = 90$ and $p = 0.5$. We approximate the probability of 60 or more correct, $P(N \geq 60)$, by $P(Z > 59.5)$ where $Z \sim N(np, np(1-p))$, i.e. $Z \sim N(45, 22.5)$. Figure 3.25 shows the calculation of $P(Z < 59.5)$ in Minitab.

So $P(Z > 59.5) = 1 - 0.9989 = 0.0011$, i.e. the probability of achieving such a good result is very small. We conclude that it is most unlikely that the student guessed as many as she claimed.

```
MTB > let k1=sqrt(22.5)
MTB > cdf 59.5;
  SUBC> normal 45 k1.
     59.5000    0.9989
```

Fig 3.25 Calculation of P(Z < 59.5).

Normal approximation to the Poisson distribution

A Poisson random variable with mean parameter μ (where μ is large, e.g. $\mu > 10$) has properties very similar to a normally distributed random variable with mean and variance both equal to μ. The approximation improves as μ increases.

Like the binomial distribution, the Poisson distribution is discrete, so a continuity correction will give a better approximation.

Example 12

Books are returned to a library at an average rate of 30 per hour. The returns are independent and random, and so, if we assume that there are no variations in the use of the library at different times of day, the number of books returned in any particular hour will have a Poisson distribution. What is the probability that fewer than 20 books are returned between 3 p.m. and 4 p.m.?

Let N be the number of books returned in the specified hour. To answer this question using the Poisson distribution, we must calculate all 20 probabilities from $P(N = 0)$ to $P(N = 19)$. Alternatively, using the Normal approximation:

$$P(N < 20) = P(X < 19.5) \text{ where } X \sim N(30, 30) = \Phi\left(\frac{19.5 - 30}{\sqrt{30}}\right) = 0.0276$$

So the odds against the library receiving fewer than 20 returns in a given hour are about 36 to 1.

EXERCISES ON 3.10

4. Two competitors A and B are involved in a general knowledge quiz. A knows the correct answer to questions with probability 0.8, whilst for B the probability is 0.3, independently of A. Out of 100 questions, which they both answer, calculate approximate probabilities for the following events.
   (a) There are between 50 and 60 questions (inclusive) where A knows the correct answer, but B doesn't.
   (b) B answers correctly at most five questions that A answers incorrectly.
   (c) At least 95 of the questions are answered correctly, either by A or by B.

   Hint: Calculate the relevant probabilities for answering one question, and then approximate the binomial distribution appropriately.
5. What is the approximate probability of between 490 and 510 heads inclusive in 1000 tosses of a fair coin?

The arithmetic of expected values (means and variances)

From an early age, children are taught to add and subtract numbers. The success of statistics in application to problems of everyday life is due to the fact that its laws state the equivalent arithmetic of random variables (i.e. when numbers are replaced by random quantities which have a probability distribution). In particular, statistics shows us how to calculate the mean and variance of, for example, the sum of two random variables given the mean and variance of the random variables in the sum. The mathematics behind this is quite deep, but an intuitive understanding, together with some very simple rules, goes a long way (sufficient for our purposes here).

Imagine you have two large collections A and B of objects whose weights (in some unspecified units) have means and variances according to the following table:

	Objects A	Objects B
Mean	10	20
Variance	16	25

Now imagine that these objects are placed in two parallel rows with the objects ordered from lightest to heaviest and that we operate with these two rows under two different experimental procedures:

Experiment 1
Choose an object from A at random – call its weight X.
Choose an object from B at random – call its weight Y.

Experiment 2
Choose one position along the length of the two rows, and then choose an object from A in that position and one from B **from the same position** – call their weights X and Y again.

Now let's answer some simple questions; in each case the questions are translated into a more practical framework, so that the answers should become apparent:

What is?	Answer	Sample equivalent
$E(X)$	10	Repeatedly sample from A and take the average
$E(2X)$	20	Repeatedly sample from A, double the value and then take the average
$E(Y)$	20	Repeatedly sample from B and take the average
$E(X+Y)$	30	Repeatedly sample A and B and take the average of the sums of each of the two weights

At this stage, all seems obvious, but pause to think whether any of these statements depends on whether we are doing Experiment 1 or 2. The answer is 'No! – even for the last one.' The expected value of a sum of random variables is *always* equal to the sum of the expected values, and the answers above can be expressed concisely by the statement that $E(aX + bY) = aE(X) + bE(Y)$ for *any* random variables X, Y, and constants a, b.

Now, suppose we consider the equivalent table for variances:

What is?	Answer	Sample equivalent
$Var(X)$	16	Repeatedly sample from A and take the sample variance
$Var(2X)$	64	Repeatedly sample from A, double the values and then take the sample variance
$Var(Y)$	25	Repeatedly sample from B and take the variance
$Var(X+Y)$ (Experiment 1)	41	Repeatedly sample A and B and take the average of the sums of each of the two weights
$Var(X+Y)$ (Experiment 2)	??	

The answers for the first and third rows of this table are obvious, and present no difficulty. The answer for the second row should cause you to do a simple experiment of calculating the sample variance for values 1, 2, 3 (first answer), and then for values 2, 4, 6 (fourth answer), to realise that doubling the values will mean the variance will increase by a factor of 4, or more generally, $Var(cX) = c^2 Var(X)$.

The crunch comes with the fourth row. Intuitively, might we not anticipate that the scatter in the sum of two random quantities, as represented by the variance, should be the sum of the individual scatters, i.e. $Var(X+Y) = Var(X) + Var(Y)$? This is indeed the case for Experiment 1. If so, is it not then obvious that in Experiment 2 it is likely that $Var(X+Y) > Var(X) + Var(Y)$? Argue intuitively as follows: in Experiment 1, when we have selected from A, we then disregard the position of this selection and select at random from B, so there is a good chance that a large weight from A is offset by a small weight from B or vice versa. This is what makes the **variances add**. For Experiment 2, however, having chosen the position, we choose from A and B **at that same position**, so that if the weight from A is small, so is that from B and similarly for large weights – thus increasing the spread in the sum of the values. In this situation the random variables X and Y are **positively correlated**. (If we chose from opposite ends they would be negatively correlated, and then we would expect the variance of the sum to be **less** than the sum of the variances.)

Hence the rule for variances is that

$$Var(X+Y) = Var(X) + Var(Y) \text{ if } X \text{ and } Y \text{ are uncorrelated}$$

Statistical independence, which we characteristically spot when observations pertain to different and unrelated samples (statistically independent samples) automatically implies uncorrelatedness, and hence in many practical situations, this simple rule for variances is applicable. As a consequence if $X_1, X_2, \ldots, X_n$ is a (statistically independent) random sample from a population with mean μ and variance σ^2 then

$$E(X_1 + X_2 + \cdots + X_n) = n\mu \quad \text{and} \quad Var(X_1 + X_2 + \cdots + X_n) = n\sigma^2$$

$$E(\bar{X}) = E\left(\frac{X_1 + X_2 + \cdots + X_n}{n}\right) = \mu \quad \text{and} \quad Var(\bar{X}) = \frac{\sigma^2}{n}$$

From this statement, because of the n in the denominator, it follows that (generally speaking) sampling and taking an average is going to give an answer nearer to the mean than if you didn't take an average (see Fig 3.26). Of course there are counter-examples, but then mathematics wouldn't be mathematics without counter-examples!

Linear combinations of normally distributed random variables

When dealing with normally distributed random variables and taking sums, the resulting distribution is always normal. If X and Y are statistically independent random variables, and $X \sim N(\mu_x, \sigma_x^2)$ and $Y \sim N(\mu_y, \sigma_y^2)$, and if a and b are constants, then

$$Z = aX + bY \sim N(a\mu_x + b\mu_y, a^2\sigma_x^2 + b^2\sigma_y^2)$$

Example 13

A businessman commits himself to two financial transactions. Let P_1 and P_2 be the profit (in pounds) on the first and second transactions respectively, and suppose that $P_1 \sim N(250, 100)$ and $P_2 \sim N(300, 400)$. (a) What is the probability that the total profit exceeds £650? (b) What is the probability that the profit in the first transaction is greater than that in the second?

(a) The total profit, $P_1 + P_2 \sim N(550, 500)$. So

$$P(P_1 + P_2 > 650) = 1 - \Phi\left(\frac{650 - 550}{\sqrt{500}}\right) = 1 - \Phi(4.47) \approx 0$$

i.e. it is most unlikely that the total profit will exceed £650.

(b) We need to calculate $P(P_1 > P_2)$. This is equivalent to $P(P_1 - P_2) > 0$. $P_1 - P_2$ is a linear combination of P_1 and P_2 (with $a = 1, b = -1$).

$$E(P_1 - P_2) = 250 - 300 = -50$$
$$Var(P_1 - P_2) = 1^2 \times Var(P_1) + (-1)^2 \times Var(P_2)$$
$$= Var(P_1) + Var(P_2) = 100 + 400 = 500$$

So

$$P_1 - P_2 \sim N(-50, 500)$$

$$P(P_1 - P_2 > 0) = 1 - \Phi\left(\frac{0 - (-50)}{\sqrt{500}}\right) = 1 - \Phi(2.236) = 0.0127$$

So there is just over a 1% chance that the profit on the first transaction will exceed that on the second.

Example 14

Suppose that $X_1, X_2, \ldots, X_{10}$ is a sample of 10 independent observations from a normal distribution with mean 200 and variance 100. How is the mean of these 10 observations distributed?

Let M be the mean of the observations,

$$M = \frac{X_1 + X_2 + \cdots + X_{10}}{10}$$

M is a linear combination of 10 random variables, so M is normally distributed. Also, according to the arithmetic of means and variances described above,

$$E(M) = \frac{1}{10}[E(X_1) + E(X_2) + \cdots + E(X_{10})]$$

$$= \frac{1}{10}(200 + 200 + \cdots + 200) = 200$$

$$Var(M) = \frac{1}{100}[Var(X_1) + Var(X_2) + \cdots + Var(X_{10})]$$

$$= \frac{1}{100}(100 + 100 + \cdots + 100) = 10$$

Figure 3.26 shows the probability density functions of (a) any of the original X_i and (b) the mean M.

We can generalise this result as follows:

If $X_1, X_2, \ldots, X_n$ are independent random variables from the distribution $N(\mu, \sigma^2)$ and if

$$\bar{X} = \frac{1}{n}\sum_{i=1}^{n} X_i$$

then

$$\bar{X} \sim N(\mu, \sigma^2/n)$$

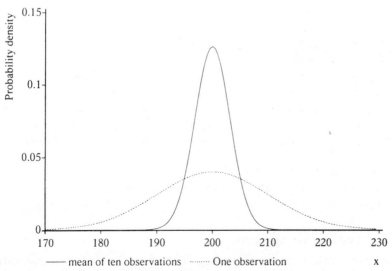

Fig 3.26 Probability density functions of original values and mean.

By a similar argument, the sum

$$S = \sum_{i=1}^{n} X_i$$

of n $N(\mu, \sigma^2)$ random variables will have the distribution $S \sim N(n\mu, n\sigma^2)$.

Example 15

A baker claims that the weight of crusty rolls is normally distributed with mean 60 g and standard deviation 3 g. If the baker's claim is correct, what is the probability that, when I weigh nine of them and calculate their mean weight, I will get a result less than 57 g?

Call the sample $X_1, X_2, \ldots, X_9$. According to the baker, each $X_i \sim N(60, 9)$, which means that $\bar{X} \sim N(60, 9/9)$.

$$P(\bar{X} < 57) = \Phi\left(\frac{57 - 60}{1}\right) = \Phi(-3) = 0.00135$$

So I am very unlikely to observe a mean less than 57 g.

Note: if I *do* obtain such a result, there are two possible causes:

1. I have been *very* unlucky.
2. The baker's claim is incorrect.

Before confronting the baker with my results, I should consider the following:

- The baker's mean of 60 may be correct, but the standard deviation may be wrong.
- The values may not be normally distributed.
- Are the observed values independent? Perhaps one batch of rolls came out a little underweight.

EXERCISE ON 3.10

6. The playing time of a particular brand of E180 video tape is normally distributed with mean 185 minutes and standard deviation 1.8 minutes. The playing time of the same brand of E240 video tape is normally distributed with mean 246 minutes and standard deviation 2.2 minutes. Let T_1 and T_2 be the playing times of an E180 and an E240, respectively, selected at random.
   (a) What is $P(4T_1 > 3T_2)$?
   (b) What is $P(T_1 + 60 > T_2)$?
   (c) What is the probability that I can record the whole of a $7\frac{1}{4}$ hour film on the two tapes?

The central limit theorem

We have seen that any linear combination of normally distributed random variables will itself have a normal distribution, with appropriate mean and variance. An

Probability Distributions

important statistical theorem, called the **central limit theorem**, states that:

> If we draw samples of size n randomly from a population whose mean is μ and variance σ^2, the sample means $\bar{X}$ are approximately distributed $N(\mu, \sigma^2/n)$ for sufficiently large n, irrespective of the shape of the population distribution.

Of course, if the population has a normal distribution, the result holds for any n. If the population is approximately normal (e.g. binomial with p near 0.5) it will be true for quite small n. For distributions which are very different in shape from the normal curve (e.g. exponential), $n > 30$ is a good rule of thumb.

Simulation

In the following simulation, we generate 40 columns of 100 random values each from the exponential distribution with mean 2.5. The columns are added to create a column of 100 sums, and then divided by 40 to create 100 means. We then plot a histogram of the first column (representing the population distribution) and a histogram of the column of means. The Minitab commands to produce the random values, draw the histograms and calculate the means are shown in Fig 3.27, and the resulting histograms in Figs 3.28 and 3.29.

The shape of the curve for the means appears to be a satisfactory approximation to the normal curve.

Recall that the mean and standard deviation of an exponential distribution are equal. The mean and standard deviation of c1 should therefore be about 2.5. The mean of c41 should also be about 2.5, and the standard deviation should be $2.5/\sqrt{40} = 0.395$. All of these values are reflected reasonably well by the observed results.

Repeat the simulation for the following distributions:

(a) binomial, $n = 20, p = 0.1$;
(b) Poisson, $\mu = 1$.

```
MTB > random 100 c1-c40;
SUBC> exponential 2.5.
MTB > add c1-c40 c41
MTB > let c41=c41/40
MTB > histogram c1
MTB > histogram c41

MTB > mean c1
     MEAN     =     2.5334
MTB > stdev c1
     ST.DEV.  =     2.3631
MTB > mean c41
     MEAN     =     2.4967
MTB > stdev c41
     ST.DEV.  =     0.43293
```

Fig 3.27 Minitab commands for the simulation.

```
Histogram of C1    N = 100

Midpoint    Count
       0     16    ****************
       1     29    *****************************
       2     14    **************
       3     13    *************
       4     11    ***********
       5      7    *******
       6      4    ****
       7      3    ***
       8      0
       9      0
      10      2    **
      11      1    *
```

Fig 3.28 Minitab histogram of the exponentially distributed values.

```
Histogram of C41   N = 100

Midpoint    Count
     1.2      1    *
     1.4      0
     1.6      2    **
     1.8      2    **
     2.0     10    **********
     2.2     19    *******************
     2.4     19    *******************
     2.6     18    ******************
     2.8     11    ***********
     3.0      8    ********
     3.2      7    *******
     3.4      2    **
     3.6      1    *
```

Fig 3.29 Minitab histogram of the means.

Example 16

The lifetimes of rubber seals in dishwashers are exponentially distributed with mean three years. A service engineer monitors 100 seals and records their lifetimes from installation to replacement. What is the probability that the mean of the engineer's 100 observations will be less than 2.5 years?

Let the life times be $T_1, T_2, \ldots, T_{100}$. Then each T_i is exponential with mean 3 and standard deviation 3 (since mean and standard deviation are equal in the exponential distribution). The mean $\bar{T}$ will be approximately normally distributed with mean 3 and standard deviation $3/\sqrt{100} = 0.3$. So

$$P(\bar{T} < 2.5) = \Phi\left(\frac{2.5 - 3}{0.3}\right) = \Phi(-1.6667) = 0.04779$$

The probability of observing a mean less than 2.5 is 0.04779 (i.e. about 1 chance in 20).

A consequence of the central limit theorem is that since $\bar{X}$ is approximately $N(\mu, \sigma^2/n)$, $S = \sum_{i=1}^{n} X_i$ will be approximately $N(n\mu, n\sigma^2)$ for large enough n, irrespective of the shape of the population distribution.

Example 17

A total of 150 schoolchildren take part in a concert which is attended by parents. For each child, the number of parents attending has the following distribution:

Number of parents	0	1	2
Probability	0.2	0.3	0.5

If the school hall will hold an audience of 210 (in addition to the 150 children), what is the probability that the concert will be oversubscribed?

Let X_i be the number of parents attending for child i.

$$E(X_i) = 0 \times 0.2 + 1 \times 0.3 + 2 \times 0.5 = 1.3$$
$$E(X_i^2) = 0^2 \times 0.2 + 1^2 \times 0.3 + 2^2 \times 0.5 = 2.3$$
$$Var(X_i) = E(X_i^2) - [E(X_i)]^2 = 2.3 - (1.3)^2 = 0.61$$

So

$$S = \sum_{i=1}^{100} X_i \sim N(150 \times 1.3, 150 \times 0.61) = N(195, 91.5)$$

$$P(S \geq 211) = 1 - \Phi\left(\frac{210.5 - 195}{\sqrt{91.5}}\right) \quad \text{(with the continuity correction)}$$

$$= 1 - \Phi(1.6204) = 0.0526$$

The probability that the concert is oversubscribed is 0.0526 (i.e. once again about 1 chance in 20).

EXERCISES ON 3.10

7. Infra-red detectors are factory set to have an average range of detection of 20 metres, with standard deviation 0.5 metres. A random sample of 100 units is tested by the quality control department of the factory. What is the approximate probability that they calculate a mean range that is less than 19.9 metres?

8. A newspaper prints 75 free advertisements every day, and allows 40 column inches for this purpose. The mean length of advertisements is 0.5 column inches, with a standard deviation of 0.3 column inches. On how many days in a

year would you expect the total length of advertisements to exceed that allocated?

3.11 Sampling distributions

A sample of size n is a set of values $x_1, x_2, \ldots, x_n$ which are realisations of the random variables $X_1, X_2, \ldots, X_n$. Any function of these random variables (e.g. $\bar{X}$), which we may design to estimate a population parameter, is called a **sample statistic** or, in short, a **statistic**. Every statistic is itself a random variable and as such it has its own probability distribution, called its **sampling distribution**. The central limit theorem tells us that, for large enough n, the sampling distribution of $\bar{X}$ is $N(\mu, \sigma^2/n)$.

Note that the word **statistic** is often used interchangeably for both the random variable (e.g. $\bar{X}$) and its realisation (e.g. $\bar{x} = (1/n) \sum_{i=1}^{n} x_i$). This leads to confusion for many students of statistics!

Knowing about sampling distributions is fundamental to two key areas of statistics: estimation and hypothesis testing.

If we use a statistic to estimate some property of a population, the sampling distribution helps us to work out how accurate our estimate is likely to be. If we have a theory (called a hypothesis) about a population characteristic (e.g. that the mean is equal to, or perhaps greater than, a particular value) we can test the validity of our theory by calculating a statistic and using our knowledge of its sampling distribution to decide whether its observed value supports or contradicts our theory. These are the issues that will concern us in Chapters 4 and 5.

Summary

If X is a continuous random variable with probability density function $f(x)$,

$$P(a < X < b) = \int_a^b f(x)dx$$

The cumulative distribution function is

$$F(x) = P(X < x) = \int_{-\infty}^{x} f(y)dy$$

Hence

$$P(a < X < b) = F(b) - F(a)$$

The mean or expected value is

$$E(X) = \int_{-\infty}^{\infty} xf(x)dx$$

Exponential distribution

If continuous random variable T is the time between events in a Poisson process with mean μ per unit time, T will have an *exponential* distribution with mean μ^{-1}.

$$f(t) = \mu e^{-\mu t}$$
$$F(t) = P(T \le t) = 1 - e^{-\mu t}$$
$$E(T) = \mu^{-1} \quad Var(T) = \mu^{-2}$$

Means and variances

For all random variables: $E(aX + b) = aE(X) + bE(Y)$.
  Also, $Var(cX) = c^2 Var(X)$.
  For uncorrelated random variables: $Var(X + Y) = Var(X) + Var(Y)$.

Normal distribution

If $X \sim N(\mu, \sigma^2)$,

$$f(x) = \frac{1}{\sqrt{2\pi\sigma^2}} \exp\left[-\frac{1}{2}\left(\frac{x-\mu}{\sigma}\right)^2\right] \quad (-\infty < x < \infty)$$
$$E(X) = \mu \quad Var(X) = \sigma^2$$
$$\frac{X - \mu}{\sigma} \sim N(0, 1)$$

The cumulative distribution function of $(X - \mu)/\sigma$, $\Phi(x)$, is found from tables or a computer package.

$N(\mu, \sigma^2)$ may be used to approximate $Bi(n, p)$ provided n is large, p is not too close to 0 or 1, $\mu = np$ and $\sigma^2 = np(1-p)$.

$N(\mu, \sigma^2)$ may be used to approximate $Po(\mu)$ provided μ is large (> 10) and $\sigma^2 = \mu$.

If X and Y are statistically independent, $X \sim N(\mu_x, \sigma_x^2)$ and $Y \sim N(\mu_y, \sigma_y^2)$, and if a and b are constants, then

$$Z = aX + bY \sim N(a\mu_x + b\mu_y, a^2\sigma_x^2 + b^2\sigma_y^2)$$

If $X_1, X_2, \ldots, X_n$ are independent random variables from $N(\mu, \sigma^2)$ then

$$\bar{X} \sim N(\mu, \sigma^2/n) \quad \text{and} \quad S = \sum_{i=1}^{n} X_i \sim N(n\mu, n\sigma^2)$$

Central limit theorem

If $X_1, X_2, \ldots, X_n$ are independent random variables from *any* distribution with mean μ and variance σ^2, *approximately* (for sufficiently large n)

$$\bar{X} \sim N(\mu, \sigma^2/n) \quad \text{and} \quad S \sim N(n\mu, n\sigma^2)$$

FURTHER EXERCISES

1. Articles from a production line have a certain proportion of defectives p. From a supposedly large population, 10 articles are chosen and tested, with acceptance if at most two are found defective. Calculate the probability $A(p)$ of

acceptance. Using values of $p = 0.01, 0.02, 0.03, \ldots, 0.20$, calculate $A(p)$, and plot a graph.

2. A two-engine aircraft can make a successful flight provided at least one engine functions throughout. For a four-engine plane, at least two engines must function throughout. If p is the probability of an engine functioning normally, show that a four-engine plane is more reliable than a two-engine plane if $p > \frac{2}{3}$.

3. The data below give the frequency of various birth patterns for children of families with at least four children. Draw up a frequency table for the number of girls in a family. Using the overall proportion of female births, compare the observed frequencies with expected frequencies assuming a binomial distribution. Compare the sample variance with that of the fitted binomial distribution.

MMMM	246	FMMM	205	MFMM	217	FFMM	207
MMMF	223	FMMF	217	MFMF	222	FFMF	204
MMFM	230	FMFM	205	MFFM	175	FFFM	182
MMFF	224	FMFF	182	MFFF	221	FFFF	183

4. The annual rainfall (in inches) in a certain region is normally distributed with mean 40 and standard deviation 4. What is the probability that starting with this year, it will take over 10 years before a year occurs having a rainfall of over 50 inches? What assumptions are you making?

5.

Number of males in litter	Size of litter			
	4	5	6	7
0	1	2	3	–
1	14	20	16	21
2	23	41	53	63
3	14	35	78	117
4	1	14	53	104
5	–	4	18	46
6	–	–	–	21
7	–	–	–	2
Totals	53	116	221	374

The table given above, containing part of data collected by Parkes from herd-book records of Duroc-Jersey pigs, shows the distribution of sex in litters of four, five, six and seven pigs. Examine whether these data are consistent with the theory that the number of males within a litter of given size is a binomial variable, the sex ratio being independent of litter size. If you are not altogether satisfied with this theory, in what direction or directions does it seem to fail?

6. The failure time distribution of an electronic component is exponential with $\mu = 1000$ hours. The company decides to offer a 'money-back' guarantee if the component fails to last at least x hours. The economics of the industry is such that each component not returned yields a profit of 10 units, whilst each returned unit results in a loss of five units. For what value of x is the overall profit of the proposed strategy five units per component sold?

7. There are four petrol stations in my town, and they vary their prices every day in such a way that I can never predict which will be the cheapest. I adopt the following strategy:
   (a) choose a station at random and visit it to check the price of a litre of petrol;
   (b) choose a second (different) station at random and, if its price is lower than the first, buy my petrol there;
   (c) if the second price is higher, visit a third station and so on until a price lower than all previous prices is found;
   (d) if the first happens to be cheapest, return to it (after visiting all four stations) and buy the petrol there.

   On a day when the four prices are 49p, 50p, 51p and 52p per litre (not necessarily in the order visited), what is the expected price per litre that I will pay?

   Hint: Consider all possible orderings of the four stations, and deduce the price paid for each ordering.

TUTORIAL PROBLEM

For this problem you will need an A3 piece of cardboard, an A3 piece of paper marked into 1 cm squares and with a centrally located pair of x, y axes and a dart. Place the paper on top of the protective cardboard on the floor. Holding the dart at shoulder height and aiming approximately for the origin of the axes, let the dart fall onto the paper. Note the x, y co-ordinates of the point that the dart strikes on a piece of paper. (You should be able to judge it accurately to about 0.1 cm by eye.) Then repeat the experiment until you have 50 x, y pairs of values. Work out, for each pair, r^2, the square of the distance from the target point $(0, 0)$. Examine the nature of the distribution of the x values, the y values and the r^2 values using histograms or other methods. Make tentative suggestions as to which of the theoretical distributions might be suitable for modelling these values. With the help of your histograms make a preliminary judgement as to whether you think that on average you are on target (as opposed to there being a bias in your aim overall).

4 • Estimation

In Chapter 1 we distinguished between the two branches of statistics: descriptive and inferential. We explored descriptive techniques in Chapter 2, and in Chapter 3 we proposed various probability distributions for use in describing or modelling datasets which occur in real life. Such statistical models usually involve one or more parameters whose values are unknown, but which can be estimated, given data. The science of statistics involves the construction of appropriate models for the description of the chaotic nature of data, methods for estimating unknown parameters in the model, the validation of the chosen model on both grounds of a priori appropriateness and also because 'it seems empirically to work', and then finally the use of the model to make predictions and assertions. In this chapter we concentrate on methods of estimating the parameters of the model.

If we toss a 'fair' coin ten times, the number of heads has a completely specified distribution (binomial, $Bi(10, p)$) where the probability of heads is $p = 0.5$. If, however, in a sample survey, 20 out of 100 people had salaries over £20000, we cannot deduce that the population (from which the sample was taken) proportion of salaries greater than £20000 *is* 0.2, because we have not sampled the whole population. Nevertheless it does seem intuitively obvious that taking the sample proportion is a 'good' method of estimating the true proportion p even though this begs the whole question of what is meant by an estimation method being 'good'. So how can we justify our intuition about estimating in this simple case, and, at the same time, develop methods of estimating for application in more complicated cases, when our intuition fails us?

It is particularly important in what follows to distinguish between the parameter of interest, the numerical value which we can calculate to estimate the parameter, and the actual method of estimation used. An analogy may help: a professional cook distinguishes between the concept of a fruit cake, a particular realisation of a fruit cake given data (the ingredients used) and the recipe employed (designed to produce a fruit cake). There are many such recipes, and some tend to produce 'better' fruit cakes than others. Similarly, there are in general lots of different methods (recipes) for estimating a parameter in a statistical model: these different methods, for the same parameter, are usually referred to as different estimators. We need to know how to construct different methods of estimation, or different estimators, and how they can be compared. This cannot be done simply by looking at the particular numerical answers these estimators produce given a particular dataset (perhaps where the true answer is known) simply because we must always take into account the essential randomness inherent in the data. Each estimator will generate its own sampling distribution which can then be used to make appropriate comparisons. So we build on our knowledge of sampling distributions, discussed at the end of Chapter 3, to begin our understanding of some of the basic principles of statistical estimation.

4.1 A simple estimation problem

How many times have you been asked to contribute to a survey or questionnaire? To keep our example of manageable scale, assume that there are just **four** members of our population, **three** of which are sampled. Suppose the individual characteristic measured has population values 1, 2, 3, 4 and we ask the question – should we use the sample mean or the sample median to estimate the population mean, known in this case to be 2.5? If we sample with replacement there are $4^3 = 64$ different samples possible, whereas if we sample without replacement there are just

$$\binom{4}{3} = 4 \text{ samples}$$

By enumerating the possible samples we can, somewhat laboriously, write down the sampling distribution of these three methods (recipes) of estimation, namely the sample mean and the sample median based on a sample of size 3 taken with replacement, and the mean of a sample of size 3 taken without replacement. (For example, in the case of sampling with replacement, the sample mean can equal 2 from six samples of the form (1, 2, 3), three samples of the form (1, 1, 4), and one sample of the form (2, 2, 2); hence the probability of the sample mean equalling 2 is $(6 + 3 + 1)/64$ equalling 0.15625.) Table 4.1 shows the results.

What do the calculations reveal? First, that all methods never get the right answer! However, on average, they all get the correct answer (as the sampling distributions all have mean values equal to 2.5). As for the spread around the true value for the sample mean distribution, as measured by the variance, this is greatest for the median estimator and least for the mean when sampling without replacement. So this indicates that of the three methods, the best is taking the mean based on a sample without replacement, and the least satisfactory is that based on the median of a sample taken with replacement.

Table 4.1 Sampling distribution for three methods of estimation from a population of values 1, 2, 3, 4.

Value	The mean sampling with replacement	The median sampling with replacement	The mean sampling without replacement
1	0.015625	0.15625	
1.33	0.046875		
1.66	0.09375		
2	0.15625	0.34375	0.25
2.33	0.1875		0.25
2.66	0.1875		0.25
3	0.15625	0.34375	0.25
3.33	0.09375		
3.66	0.046875		
4	0.015625	0.15625	
Mean	2.5	2.5	2.5
Variance	0.4167	0.875	0.13888

The first property, that of being 'right on average', is called **unbiasedness**. An **estimator of a parameter** is said to be **unbiased** for that parameter, if the **sampling distribution of the estimator has a theoretical mean equal to the parameter**. For two such estimators, the one with **smaller variance** is regarded as superior.

EXERCISES ON 4.1

1. Work out the distribution of the median of a sample of size 3 taken without replacement from the population 1, 2, 3, 4 as above.
2. Produce an equivalent table to Table 4.1 assuming that the population values are 1, 2, 3, 8. Show that in this case, the sample median is no longer an unbiased estimator of the population mean. (Does this contradict the common advice to use the median to measure central tendency when a sample contains an 'outlier'?)
3. A coin with unknown probability of heads p is tossed 20 times. As a first estimator of p, consider the sample proportion of heads in the 20 tosses, and as a second estimator of p, consider the proportion of heads in the sequence of odd-numbered tosses. Which do you think is the better estimator and why?
4. A coin with unknown probability of heads p is tossed 20 times and the sequence of results is TTTTTHHHTTHHTHTTTHTH. The proportion of heads is therefore 0.4. Counting the number of tosses up to and including the first of each heads gives the numbers 6 1 1 3 1 2 4 2 giving an average run length of 2.5. Take the reciprocal of this average (which just happens to be 0.4 again) to be another estimate of p. (Why does this make sense? Refer to the geometric distribution in Chapter 3.) This experiment is repeated 50 times and the sample of pairs of estimates of p is given below (with the first of each row being the observed proportion of heads in each 20 tosses, and the second being the reciprocal of the average run length to each heads). Calculate sample means and variances for the empirical distributions from these two methods of estimation, and also sketch comparative histograms. Given the information that the actual value of p is 0.4, which method of estimation do you think is preferable and why?

```
0.4    0.35  0.3   0.3   0.4   0.45  0.35  0.35  0.4   0.4
0.5    0.58  0.32  0.33  0.4   0.45  0.37  0.54  0.47  0.4

0.3    0.4   0.25  0.4   0.45  0.4   0.7   0.2   0.35  0.4
0.33   0.4   0.33  0.44  0.47  0.4   0.74  0.25  0.35  0.47

0.45   0.45  0.45  0.6   0.55  0.3   0.45  0.4   0.5   0.6
0.56   0.47  0.45  0.6   0.61  0.32  0.64  0.42  0.59  0.71

0.45   0.45  0.25  0.45  0.45  0.35  0.35  0.25  0.65  0.3
0.45   0.45  0.25  0.45  0.5   0.35  0.41  0.28  0.65  0.35

0.25   0.45  0.5   0.45  0.5   0.3   0.35  0.35  0.4   0.4
0.28   0.45  0.5   0.5   0.5   0.3   0.37  0.37  0.42  0.4
```

4.2 Maximum likelihood estimation

The new method of estimating the parameter p in a sequence of independent Bernoulli trials introduced in Exercise 4 in Section 4.1 is not used because it is biased, at least for small samples such as 20. The science of statistical estimation is as important now (when statisticians develop, for example, complex models to estimate the amount of oil left in an underground reservoir) as it was for the early pioneers of statistics who were concerned with experimental errors in astronomical observations. Hence any general method of estimation that can be used with confidence in standard situations and which will also provide a starting point for estimation in new situations has got to be important.

The **method of maximum likelihood estimation** plays such a fundamental role in the science of statistical estimation, and since the coin-tossing model is such a familiar model to us, it is convenient to use it once again to introduce this method.

Suppose that out of 10 trials, four are successful, thus giving a sample estimate of p equal to 4/10. Elementary probability theory implies that the probability of obtaining four successes in 10 trials is given by:

$$P(4 \text{ out of } 10 \text{ are successes}) = \binom{10}{4} p^4 (1-p)^6$$

Regarded as **a function of p**, the unknown parameter, rather than as a function of the data values (four successes in 10 trials) this is called the **likelihood function** as it represents (as p varies) the likelihood of what we know to have occurred. If we evaluate this likelihood function with $p = 0.1$ and with $p = 0.5$, then it is easy to show that the two resulting likelihood values are in a ratio of about 18 to 1 (in favour of $p = 0.5$). This suggests that a value of $p = 0.5$ is a better choice than $p = 0.1$ **given the data**. Following this principle to its logical conclusion suggests **taking that value of p that maximises the likelihood function**. Simple differentiation shows that this occurs when $p = 4/10$. Consequently $p = 0.4$ is the maximum likelihood estimate of p in this case. We write $\hat{p} = 0.4$.

We can easily repeat this simple argument in the general case where we know r out of n trials are 'success', by differentiating the function $p^r(1-p)^{n-r}$ and equating the derivative with respect to p to zero. (Note that the combinatorial term, because it does not depend on p, will not affect the calculation of the maximum likelihood estimator.)

$$\frac{d}{dp} p^r (1-p)^{n-r} = rp^{r-1}(1-p)^{n-r} - p^r(n-r)(1-p)^{n-r-1}$$

$$= 0 \text{ when } r(1-p) - (n-r)p = 0, \text{ i.e. } p = \frac{r}{n}$$

Figure 4.1 shows graphs of the likelihood function for $r = 2, 5, 8$ and $n = 10$. In each case it is obvious that the maximum is achieved when $p = 2/10$, $p = 5/10$ or $p = 8/10$, respectively.

Thus, whenever a binomial model applies to data, we can estimate the parameter p by the maximum likelihood estimate $\hat{p} = r/n$. Of course, this estimate is not the correct value of p – just an estimate. How accurate is the estimate? Sensible though this question seems to be, the answer is inevitably that if we could determine how accurate the estimate is, then we must know the true value

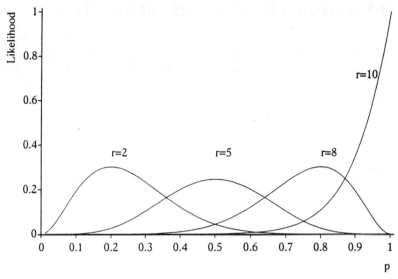

Fig 4.1 The likelihood function for $r = 2, 5, 8$ and $n = 10$.

of p and hence we do not need to estimate in the first place! However, if we look at the **method of estimation**, and use information concerning the distribution of the associated random variable, then we can gain information on how accurate the method of estimation is.

Return to our initial example, where we know that 4 successes occurred in 10 trials. The value 4 is the value of the random variable N equal to the total number of successes in the 10 trials. It is this random variable that has a binomial distribution $Bi(10, p)$. So we can now ask the question: what is the probability of obtaining a result at least as extreme as 4 given a value p? The value 4 can be extreme in two ways: either by being very large, considering the value of p, or by being very small. Hence we investigate how the probabilities $P(N \geq 4)$ and $P(N \leq 4)$ vary as p varies. This is easy to do using a computer program and Fig 4.2 graphs these two probabilities as p varies.

As a result, we see that if $p \leq 0.122$ then $P(N \geq 4) \leq 0.025$ and also if $p \geq 0.737$ then $P(N \leq 4) \leq 0.025$. The derived interval $(0.122, 0.737)$ is called a 95% **confidence interval** for p. By this, we do not mean that p lies in the interval $(0.122, 0.737)$ with probability 0.95 because we consider p to be a fixed constant, so that it either does or does not belong to this interval. Rather, it is the case that if this method of constructing a 95% confidence interval were to be repeated a large number of times, approximately 95% of the derived intervals would contain the true value of p.

If a sample of observations provides an estimate $\hat{\mu}$ of a parameter μ, then solving the equation $P(\text{sample estimate} \geq \hat{\mu}) = 0.025$ yields a value of the parameter μ_L, and solving the equation $P(\text{sample estimate} \leq \hat{\mu}) = 0.025$ gives a value of the parameter μ_H, then the interval (μ_L, μ_H) is a 95% confidence interval for the parameter μ. If the value 0.025 in these equations is replaced by 0.5α, then the result is a $100(1 - \alpha)\%$ confidence interval.

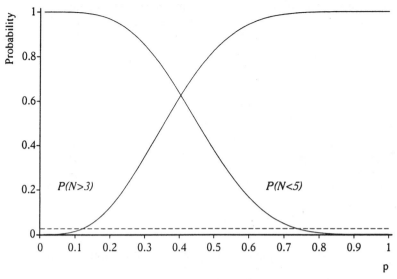

Fig 4.2 $P(N \geq 4)$ and $P(N \leq 4)$ as p varies.

EXERCISES ON 4.2

1. If there are $n = 2$ trials and $r = 1$ is the number of successes, by using the probabilities of the binomial distribution $Bi(2, p)$, show that the 95% confidence interval for p is of the form $1 - \sqrt{0.975} \leq p \leq \sqrt{0.975}$. Similarly, determine a 90% confidence interval.

2. The probability that a blood sample taken from a person chosen at random in a certain population contains a particular antibody is p. Out of 100 people tested, 38 samples had the particular antibody present. Use the normal approximation to the binomial distribution (with continuity correction) to calculate a 95% confidence interval for the true value of p.

   Hint: In this case the random variable N has the distribution $Bi(100, p)$ which may be approximated by a normal distribution $N(100p, 100p(1-p))$. Be prepared to solve quadratic equations in p.

An approximate formula for a confidence interval for p

Suppose we want to estimate an unknown proportion p on the basis of n trials. We calculate the observed proportion $\hat{p} = N/n$ in the n trials where N has the binomial distribution $Bi(n, p)$. We now approximate the distribution of $\hat{p}$ by a normal distribution with mean p and variance $p(1-p)/n$. To find a confidence interval for p we must solve the equations

$$\hat{p} = p \pm 1.96 \sqrt{\frac{p(1-p)}{n}}$$

This is complicated because of the quadratic expression in p within the square root. However, since $p(1-p) \leq \frac{1}{4}$ (check this by differentiating or completing the square), and also since the value of $p(1-p)$ doesn't change too rapidly for values of p near 0.5, we can approximate the confidence interval, making it larger if anything, by replacing $p(1-p)$ by its maximum value. So we need only solve the equations

$$\hat{p} = p \pm 1.96\sqrt{\frac{1}{4n}}$$

which (approximating 1.96 by 2) gives a 95% confidence interval for p of the form $\hat{p} \pm 1/\sqrt{n}$.

EXERCISE ON 4.2

3. A coin is tossed 10000 times and the number of heads is 5110. Using the approximate confidence interval given above, show that a 95% confidence interval for p does not include 0.5.

Returning to the principle of maximum likelihood estimation, in order to apply it in a different context, we must take the following steps:

(1) state a probability model for the data involving the unknown parameter;
(2) construct a likelihood function using the data;
(3) find the value of the parameter that maximises the likelihood function;
(4) find an appropriate confidence interval for the parameter by examining the probability of the sample estimate being extreme (either way) as the parameter varies.

By far the most difficult step is (1) – determining an appropriate model – as this involves the choice of an appropriate distribution to reflect the randomness inherent in the data. Developing a good choice depends on looking at the data and the context behind the data, and a good knowledge of common statistical models that are available. Further details of step (2) are revealed in subsequent examples.

4.3 Maximum likelihood estimation using a frequency table

Out of 42 calls made on a BT phonecard, the times in seconds have the frequency table given in Table 4.2. It is thought that such times can be modelled successfully by the use of an exponential distribution where probabilities are given by integrating the probability density function

$$f(t) = \frac{1}{\mu} e^{-t/\mu}$$

We need to estimate the mean parameter μ and consider whether the model appears to be appropriate.

One of the problems with such a frequency table (which may be all you have to go on if you did not collect the data yourself) is how to cope with the open-endedness of the last interval. In order to simplify the algebra, we will make the

Table 4.2 Lengths of calls (in seconds) of 42 telephone calls on a BT phonecard.

Time interval	Frequency	Expected frequency
0–20	10	9.6
20–40	9	7.4
40–60	7	5.7
60–80	2	4.4
80–100	1	3.4
> 100	13	11.5

problem even more pronounced by assuming that we know only the frequencies 10, 9 and 23, i.e. we only know that there are 23 observations more than 40.

To evaluate the likelihood, we have to associate with each of the 42 observations a probability – in this case, the probability of the observation falling into the class interval it belongs to. Thus we first integrate the probability density function $f(t)$ over a general interval (a, b) to obtain the probability associated with that interval equal to $e^{-a/\mu} - e^{-b/\mu}$. The probability that a call has length between 0 and 20, and thus falls into the first time interval, is given by this formula with $a = 0$ and $b = 20$ and hence there are 10 such terms in the likelihood – one for each of the 10 observations in the first time interval. Similarly, each of the nine calls in the second interval contributes a term of the same form with $a = 20$ and $b = 40$, and so on. For the last interval (the > 40 interval) the required contribution can be derived by putting $a = 40$, and letting b increase without limit to get a contribution of the form $e^{-40/\mu}$. The resulting likelihood is therefore of the form $(1 - e^{-20/\mu})^{10}(e^{-20/\mu} - e^{-40/\mu})^9 (e^{-40/\mu})^{23}$.

We can simplify this formula considerably if we replace $e^{-20/\mu}$ by p to get $(1 - p)^{10}[p(1 - p)]^9 (p^2)^{23}$ which we can reduce to $(1 - p)^{10+9} p^{9+2\times 23}$. This is of the same form as the previous likelihood for a binomial likelihood model! So differentiating and equating to zero gives the maximum likelihood estimate for p of the form $p = 55/(55 + 19)$ (put $r = 9 + 46 = 55$, and $n - r = 10 + 9 = 19$). Solving for μ gives $\mu = -20/\log(55/74) = 67.4$.

What have we gained by the use of this method? First, the answer clearly takes into account the open-endedness of the last class interval. (If we were to assume that all the 23 largest observations were in an interval of length 20, then we would get a sample mean of 36.19 – a lot smaller than the maximum likelihood estimate just produced.) With the estimated mean, we can now calculate probabilities for each interval, and hence calculate expected frequencies. For example, for the second interval, the expected frequency is given by $42(e^{-20/\mu} - e^{-40/\mu}) = 8.01$ and the table of expected frequencies (based on using the full table – see Exercise 5 at the end of this section) shows a modest success in modelling the data succinctly.

Second, because the form of the likelihood is that of a binomial likelihood, we can find a 95% confidence interval for the parameter p (and hence for μ) by the method indicated in Section 4.2. We calculate the 95% confidence interval for p (based on $r = 55$ and $n = 74$) from probabilities associated with the binomial distribution and obtain $(0.6285, 0.8378)$. Hence we obtain a 95% confidence interval for μ by solving for μ in the equation $p = e^{-20/\mu}$ to get (43, 113). The fact that this interval is so large reflects the loss of information which results when we group the data into a small number of intervals.)

Finally, this example holds the key to the application of maximum likelihood estimation to a dataset to be modelled by a continuous distribution. If, instead of a frequency table, we had 42 actual sample values, what would be the contribution to the likelihood from an observation x_1? Imagine that this observation belongs to an interval (a, b) where a and b slowly converge towards the value x_1. The probability $e^{-a/\mu} - e^{-b/\mu}$ would then become proportional to the probability density function evaluated at the point x_1, i.e.

$$\frac{1}{\mu}\exp(-x_1/\mu)$$

To summarise, the construction of the likelihood function proceeds by calculating a product of terms, one for each observation in the sample. An individual term in the product is either the probability of that observation occurring (when this is non-zero) or the probability density function evaluated at the observation. Each term will generally involve the unknown parameter(s). This step-by-step approach is displayed in Fig 4.3.

Note: It is convenient in many situations to construct the log-likelihood function (by taking the logarithm of the likelihood) instead of the likelihood. This is because the logarithm of a product is the **sum** of the logarithms of each individual term in the likelihood function, and it is easier to deal with sums than with products. Maximising the log-likelihood function will give the same answer as maximising the likelihood function – see Exercise 1 at the end of this section). The construction

Sample of Observations				
Observation 1	Observation 2 ...	... Observation i ...	... Observation n
x_1	x_2 ...	... x_i ...	... x_n
$p_1(\mu)$	$p_2(\mu)$ ...	... $p_i(\mu)$ ...	... $p_n(\mu)$
Likelihood function $= p_1(\mu)p_2(\mu)\ldots p_i(\mu)\ldots p_n(\mu)$			
Log-likelihood function $= \log p_1(\mu) + \log p_2(\mu) + \cdots + \log p_i(\mu) + \cdots + \log p_n(\mu)$			
where $p_i(\mu) = P(\text{Observation } x_i)$ if this is greater than 0			
otherwise $p_i(\mu)$ is the probability density function evaluated at x_i.			

Fig 4.3 Construction of the likelihood and log-likelihood functions for an unknown parameter μ.

Estimation 91

of the log-likelihood function for a sample of n statistically independent observations is also outlined in Fig 4.3.

With this scheme, we know how maximum likelihood estimators may be calculated in a routine way for data from the common discrete and continuous distributions. This we do in the next section. The following exercises are designed to give you practice in this methodology, all within the context of 'tossing coins' or equivalent scenarios. The first of them again illustrates how the method of maximum likelihood estimation can even be applied when you have no data! Just replace the unknown data by some algebraic constants. When you do this, you must remember that the function to be maximised is the likelihood or log-likelihood **as the parameter varies**.

EXERCISES ON 4.3

1. Show that the log-likelihood function for the binomial model based on n trials, r of which are successes, is given by $l(p) = r \log p + (n - r) \log(1 - p)$. By differentiating this function with respect to p, derive the same maximum likelihood estimate as before.

2. A coin with probability of heads p is tossed until the first heads occurs and the number of tosses is recorded (including the heads). This is repeated five times to give the number of tosses in each case equal to 2, 5, 3, 4, 12. Using the geometric distribution to model the data (why?), show that the maximum likelihood estimate of p is given by

$$\hat{p} = \frac{5}{2 + 5 + 3 + 4 + 12}$$

Explain why this makes sense intuitively (what is the sample proportion of heads in all the tosses?).

3. Lightbulbs from a production line have an unknown proportion p of defective bulbs. They are combined at random into packets containing two bulbs and bought by customers in a supermarket for a relatively small price. Assuming that a customer will only bother to complain if both bulbs in a packet are defective, and given that of 100 purchases, 25 packets are returned, estimate the proportion p using the method of maximum likelihood, and calculate a 95% confidence interval for p. What is the general formula for the maximum likelihood estimate in the case when r out of n customers return their doubly defective packets?

4. Following from Exercise 3 above, suppose now that a packet is returned if it contains at least one defective bulb. What is the maximum likelihood estimate of p then?

5. With the example of the BT phone call times (Table 4.2), repeat the construction of the likelihood function and find the maximum likelihood estimate of μ and a 95% confidence interval for μ using the full frequency table. Derive the expected frequencies quoted in Table 4.2.

4.4 Maximum likelihood estimation for common probability distributions

The Poisson distribution

Observations from a Poisson distribution with mean μ take integer values $0, 1, 2, \ldots$ with probabilities given by

$$p_x = e^{-\mu} \frac{\mu^x}{x!} \quad (x = 0, 1, 2, \ldots)$$

Taking the natural logarithm of this gives $\log p_x = -\mu + x \log \mu - \log x!$. The last term can be ignored in the calculation of the likelihood, as it does not involve the unknown parameter μ. For a sample of size n, the log-likelihood is a sum of such terms with x replaced by each sample value, hence we derive the log-likelihood function $l(\mu)$ given by

$$l(\mu) = -n\mu + \log \mu \sum_{i=1}^{n} x_i$$

(Note how this function depends on the data only through the sum of the data items – the actual values are not required. For this example, the sum of the data is said to be a **sufficient statistic** for μ. This is an important concept in statistics but beyond the scope of this book.)

Differentiating with respect to μ and equating to zero gives the equation

$$-n + \sum_{i=1}^{n} \frac{x_i}{\mu} = 0$$

with solution $\hat{\mu} = \bar{x}$.

Since the sum of n observations having a Poisson distribution with mean μ has a Poisson distribution with mean $n\mu$, confidence intervals for the true value of μ can be found by evaluating Poisson probabilities.

Example I

Consider the following data from a paper by Sykes and Temple in the *British Medical Journal* (Vol. 305), reporting on the number of acute episodes per week of asthma in a Welsh coal-mining village. For 55 weeks 30 August 1989 to 26 September 1990, the weekly figures were as follows:

3 1 7 3 6 6 0 2 9 2 2 5 5 9 5 6 0 2 0 2 5 3 10 7 9 2 2 3
2 5 11 0 6 2 3 6 11 2 2 6 8 7 3 3 9 0 7 2 4 1 3 3 4 9 6

The sample mean equals 4.38, and the sum of the observations equals 241. To find a 95% confidence interval for the mean μ, we need to solve $P(Y_1 + \cdots + Y_{55} \leq 241)$ and the equivalent probability with the inequality reversed. Equivalently, we solve:

$$\sum_{n=241}^{\infty} e^{-55\mu} \frac{(55\mu)^n}{n!} = 0.025 \quad \text{and} \quad \sum_{n=0}^{241} e^{-55\mu} \frac{(55\mu)^n}{n!} = 0.025$$

Estimation

This is no mean computational task because of the large powers and factorials required but the answer to three decimal places is (3.846, 4.971). Alternatively, the central limit theorem rises to the challenge, by allowing us to approximate the Poisson distribution with mean 55μ by a normal distribution with mean 55μ and variance 55μ. Thus the probability of a value at least 241 is approximated by the area under the standard normal probability density curve to the right of $(241 - 0.5 - 55\mu)/\sqrt{55\mu}$. Hence to find a lower confidence limit for μ, solve the equation $(240.5 - 55\mu)/\sqrt{55\mu} = 1.96$. Similarly, to find the upper limit for μ, solve the equation $(241.5 - 55\mu)/\sqrt{55\mu} = -1.96$. Each of these equations involves solving a quadratic equation in $\sqrt{\mu}$ to give the 95% confidence interval (3.87, 4.98) agreeing with the direct calculations.

EXERCISES ON 4.4

1. Incoming calls at a telephone exchange are such that the number N of calls in a time period of x seconds has a Poisson distribution with mean μx. Given that in a period of 10 seconds only one call is received, derive the equations $P(N \geq 1) = 0.025$ and $P(N \leq 1) = 0.025$ required to obtain the left- and right-hand points of the 95% confidence interval for μ. Hence show that the confidence interval is (0.00253, 0.557). (Note that the second equation has to be solved numerically!)
2. If (as in Exercise 1 above) the telephone exchange received 52 calls in one minute, use the normal approximation to the Poisson distribution $Po(60\mu)$ to find an approximate 95% confidence interval for μ.

The exponential distribution

Observations from an exponential distribution have a probability density function $f(x)$ given by

$$f(x) = \frac{1}{\mu} e^{-x/\mu}$$

for positive x values. Following the sequence of steps outlined in Fig 4.3 (taking logarithms, summing over contributions from different x values) the log-likelihood function is given by

$$l(\mu) = -n \log \mu - \sum_{i=1}^{n} \frac{x_i}{\mu}$$

(Note again that the sum of the x data is all that you need to evaluate the log-likelihood function, so it is a sufficient statistic for μ as it was for the mean of a Poisson distribution.)

By differentiating with respect to μ and equating to zero, we find that the maximum likelihood estimate of the parameter μ is equal to the sample mean $\bar{x}$. This once again is reassuring, as the parameter μ for the Poisson and exponential models as specified here is the theoretical mean of the distribution.

For example, the telephone call lengths summarised in Table 4.2 were known exactly, and their sum was equal to 3345.2 (42 calls in all). Hence the sample mean is equal to 79.65. This is of course likely to be a much better estimate of the mean parameter μ because the actual data values have been used. In order to derive a confidence interval, we have to look at the sampling distribution of the sample mean $\bar{Y}$, where each of the 42 contributions to $\bar{Y}$ have mean μ and variance μ^2. Hence, by the central limit theorem, the distribution of $\bar{Y}$ may be approximated by a normal distribution with mean μ and variance μ^2/n.

Hence

$$P(\bar{Y} \geq 79.65) = P\left(\frac{\bar{Y} - \mu}{\mu/\sqrt{42}} \geq \frac{79.65 - \mu}{\mu/\sqrt{42}}\right) = 0.025$$

i.e.

$$\frac{79.65 - \mu}{\mu/\sqrt{42}} = 1.96 \quad \text{and} \quad \mu = \frac{79.65}{1 + 1.96/\sqrt{42}} = 61.15$$

Similarly

$$P(\bar{Y} \leq 79.65) \quad \text{when} \quad \mu = \frac{79.65}{1 - 1.96/\sqrt{42}} = 114.18$$

and hence the 95% confidence interval for μ is (61.15, 114.18).

EXERCISE ON 4.4

3. Ten students were given an arithmetic task to do and were timed in seconds. They only had 120 seconds to complete it and one student didn't manage it – his time is recorded as 120* (such observations are said to be censored). Assuming that the time taken by a typical student has an exponential distribution with mean μ, write down the log-likelihood function for the sample by applying the methodology of Fig 4.3 (don't forget that the censored value has positive probability of equalling 120). Hence evaluate the maximum likelihood estimate for μ. The recorded data are:

21.7   107.0   120*   24.5   89.1   36.6   22.3   71.2   80.55   111.9

If the censored value was actually 240.1 calculate the maximum likelihood estimate of μ and a 95% confidence interval for it.

4.5 Maximum likelihood for normally distributed data

The single-sample case

A sample of size n from a normal distribution with mean μ and variance σ^2 is perhaps the most common model applied to single samples. Its frequent use is often justified as follows. Each of the observations $Y_1, Y_2, \ldots, Y_n$ can be regarded as equalling the population mean μ **plus** an error term. Each error term may be envisaged as a sum of lots of small errors from many different (unknown) sources. Consequently, by the central limit theorem, we can expect each error term to have a normal distribution with zero mean and some variance σ^2. It is useful to write this model down formally, as it is the simplest of models used in experimental design. The model is $Y_i = \mu + E_i$ where $i = 1, 2, \ldots, n$ and $E_i \sim N(0, \sigma^2)$.

Since the parameter μ is the mean of the random observations Y_i, it is natural to consider estimating μ by taking the sample mean $\bar{y}$. We therefore hope that the principle of maximum likelihood estimation will agree with this. The random observation Y_i has a normal distribution with mean μ and variance σ^2. The probability density function evaluated at the point y_i is therefore equal to

$$\frac{1}{\sqrt{2\pi}\sigma} \exp\left(\frac{-(y_i - \mu)^2}{2\sigma^2}\right)$$

and, hence, we can obtain the log-likelihood function by taking logarithms and summing all such terms corresponding to each observation, giving the result

$$-\sum_{i=1}^{n}\left(\frac{(y_i - \mu)^2}{2\sigma^2}\right) - 0.5n\log(2\pi\sigma)$$

This is of course a function of two unknown parameters, μ and σ^2, but the last term depends only on σ^2 and the first term involves μ through a sum of squares with an attached minus sign. So to maximise the log-likelihood function as a function of μ, we must **minimise the sum of squares**

$$\sum_{i=1}^{n}(y_i - \mu)^2$$

Thus the maximum likelihood estimate will be the same as the so-called **least-squares estimate** (see the method of least squares in Section 4.6), namely the value of μ which minimises this sum of squares. We can derive this easily by differentiating and equating to zero – or, as will be useful later, by using the equality

$$\sum_{i=1}^{n}(y_i - \mu)^2 = n(\bar{y} - \mu)^2 + \sum_{i=1}^{n}(y_i - \bar{y})^2$$

from which it is obvious that the least-squares estimate is obtained by choosing μ to equal $\bar{y}$, thus making the last term in the above equation zero; hence the maximum likelihood estimate is given by $\hat{\mu} = \bar{y}$.

EXERCISE ON 4.5

1. Derive the above equality by expanding the square bracket $[(y_i - \bar{y}) + (\bar{y} - \mu)]^2 = (a + b)^2$ and summing, remembering that

$$\sum_{i=1}^{n}(y_i - \bar{y}) = 0$$

Example 2

Packets of sweets are known to have a normal distribution with standard deviation 10 g. Two packets weigh 510 g and 520 g, respectively. Construct the likelihood function and the log-likelihood function and hence show that the maximum likelihood estimate of the population mean weight μ is 515 g.

The likelihood for this problem is the product of two normal density functions evaluated at the two weights, and with variance parameter 100:

$$\frac{1}{10\sqrt{2\pi}}\exp\left(-\frac{1}{200}(510-\mu)^2\right) \times \frac{1}{10\sqrt{2\pi}}\exp\left(-\frac{1}{200}(520-\mu)^2\right)$$

$$=\frac{1}{200\pi}\exp\left(-\frac{1}{200}[(510-\mu)^2+(520-\mu)^2]\right)$$

Hence the log-likelihood function is a **sum of squares**:

$$l(\mu) = -[(510-\mu)^2+(520-\mu)^2] - \ln(200\pi)$$

as the last constant (not involving μ) may be ignored. By differentiating with respect to μ, or by using the equality proved in Exercise 1 above, this sum of squares is minimised when $\mu = 1030/2$ which equals the sample mean, $\bar{y} = 515\,\text{g}$. To calculate a 95% confidence interval for μ, we must solve for μ the two probability statements

$$P(\bar{Y} \geq 515) = 0.025 \quad\text{and}\quad P(\bar{Y} \leq 515) = 0.025$$

Since $\bar{Y}$ has the normal distribution $N(\mu, 100/2)$,

$$P(\bar{Y} \geq 515) = 0.025$$

if

$$\mu = 515 + 1.96\frac{10}{\sqrt{2}}$$

and similarly for the other probability; hence the 95% confidence interval takes the form

$$515 \pm 1.96\frac{10}{\sqrt{2}}$$

This result may be generalised and stated as follows:

> If the sample mean from a sample of size n from a normal distribution with known variance σ^2 is $\bar{y}$, then a 95% confidence interval for the mean parameter μ takes the form
>
> $$\bar{y} \pm 1.96\frac{\sigma}{\sqrt{n}}$$

Note that for a $100(1-\alpha)$% confidence interval, the value 1.96 is replaced by the $100(1-\alpha/2)$ percentile of the standard normal distribution. Also note that if σ is unknown but n is large (say > 30) then σ may be replaced by the sample standard deviation s.

EXERCISE ON 4.5

2. In a sample of 50 weights of bags of sugar supposed to equal 1 kg, the mean weight was 0.95 kg, with a standard deviation of 0.14 kg. Calculate a 95% confidence interval for the population mean weight of bags of sugar and show that this does not include the supposed value 1 kg. Does your calculation of the confidence interval assume that the individual weights are normally distributed?

For what value of the standard deviation would the 95% confidence interval for the mean weight of the bags just include 1 kg?

4.6 The method of least squares

Example 2 is perhaps the simplest application of the method of least-squares estimation, which we derived from applying the principle of maximum likelihood assuming that the data was normally distributed. The method of least squares, as applied to that example, may be presented in the following concise way:

(1) construct a three-column table with the first column equal to the data, the second column equal to the expected value for each data item, and the third column the square of the difference of the first two columns;
(2) sum the third column and estimate the unknown parameter by the value that minimises this sum of squares.

This is illustrated in Table 4.3.

Table 4.3 The method of least squares.

Observation	Expected value	Square of difference
510	μ	$(510 - \mu)^2$
520	μ	$(520 - \mu)^2$
		Sum of squares
		$S(\mu) = (510 - \mu)^2 + (520 - \mu)^2$
		Least-squares value is
		$\hat{\mu} = 515 =$ sample mean

This method can be used to great effect in many different situations. The form of the expected values in the second column will depend on the model proposed and will generally involve at least one unknown parameter. Any unknown parameters are then estimated by choosing the values that minimise the sum of squares of observations minus expected values. The following examples are designed to give the reader more familiarity with this approach.

Example 3

Two sample 1 g tablets from each of two manufacturers of vitamin C are taken and the amount of vitamin C in each tablet measured. The results were 0.95, 0.83 for Manufacturer X and 0.98, 0.94 for Manufacturer Y. Estimate using the principle of least squares the two population means μ_x, μ_y.

Obviously, at the end of the day, we know we are going to estimate the two population means by the equivalent sample means, but it is nevertheless instructive to lay out the estimation procedure as an application of the principle of least squares.

Observation	Expected value	Square of difference
0.95	μ_x	$(0.95 - \mu_x)^2$
0.83	μ_x	$(0.83 - \mu_x)^2$
0.98	μ_y	$(0.98 - \mu_y)^2$
0.94	μ_y	$(0.94 - \mu_y)^2$
		Sum of squares $S(\mu_x, \mu_y) =$ $(0.95 - \mu_x)^2 + (0.83 - \mu_x)^2 +$ $(0.98 - \mu_y)^2 + (0.94 - \mu_y)^2$ Least-squares occurs when $\hat{\mu}_x = 0.89 \quad \hat{\mu}_y = 0.96$

Note that the easiest way to see this result is in the general context of a set of x data $x_1, x_2, \ldots, x_m$ and a set of y data $y_1, y_2, \ldots, y_n$. Then the expression for $S(\mu_x, \mu_y)$ is given by

$$S(\mu_x, \mu_y) = \sum_{i=1}^{m}(x_i - \mu_x)^2 + \sum_{i=1}^{n}(y_i - \mu_y)^2$$

$$= m(\bar{x} - \mu_x)^2 + n(\bar{y} - \mu_y)^2 + \sum_{i=1}^{m}(x_i - \bar{x})^2 + \sum_{i=1}^{n}(y_i - \bar{y})^2$$

by a double application of the equality proved in Exercise 1 in Section 4.5. Clearly if $\mu_x \neq \bar{x}$ and/or $\mu_y \neq \bar{y}$ then $S(\mu_x, \mu_y)$ is bigger than it need be. Hence the two population means must, by the principle of least squares, be estimated by their sample means. The important aspect of this, however, is to note that the least-squares value achieved is given by what is called the **within-groups sums of squares**

$$\sum_{i=1}^{m}(x_i - \bar{x})^2 + \sum_{i=1}^{n}(y_i - \bar{y})^2$$

and it is this quantity which is used to estimate the variance **if it is assumed that the two samples come from two populations with the same variance**.

Example 4

Two plants are grown under identical conditions and the height after one month of the first plant is 10 cm and the height of the second plant after two months is 15 cm. If biological theory suggests that the expected height of such a plant after x months of growth is bx^2 cm use the principle of least squares to estimate the unknown parameter b.

The required table for applying the method of least squares is

Observation	Expected value	Square of difference
10	b	$(10 - b)^2$
15	$4b$	$(15 - 4b)^2$
		Sum of squares $S(b) = (10 - b)^2 + (15 - 4b)^2$ Least-squares occurs when $b = 140/34 = 4.12$

The expected value of a plant's height depends on two values: the unknown parameter b and the time variable. This is a very simple example of a regression problem, and the solution may be derived by differentiating the sum of squares with respect to the unknown parameter (derivative $= S'(b) = -20 + 2b - 120 + 32b$ which equals 0 when $b = 140/34$).

Example 5

From a road atlas n pairs of towns in England and Wales are chosen at random. The distance between pair i in miles by road is Y_i and the distance on the atlas between the towns 'as the crow flies' measured in millimetres is x_i. Since for zero x value we would expect an actual distance of zero, a model that says $E(Y_i) = bx_i$ seems appropriate. (We would not expect Y_i to equal bx_i exactly as roads in England and Wales twist and turn somewhat, more so in some places than in others.) Show how you would analyse this problem using the method of least squares.

The method of least squares would lead to a table such as the one below where the dots in the fourth line of the table indicate that data values for cases $3, 4, \ldots, (n-1)$ have been missed out.

Observation	Expected value	Square of difference
Y_1	bx_1	$(Y_1 - bx_1)^2$
Y_2	bx_2	$(Y_2 - bx_2)^2$
$\vdots$	$\vdots$	$\vdots$
Y_n	bx_n	$(Y_n - bx_n)^2$
		Sum of squares
		$S(b) = \sum_{i=1}^{n}(Y_i - bx_i)^2$
		Least-squares occurs when
		$b = \sum_{i=1}^{n} Y_i x_i / \sum_{i=1}^{n} x_i^2$

EXERCISES ON 4.6

1. Prove that the minimum of $\sum_{i=1}^{n}(Y_i - bx_i)^2$ occurs when $b = \sum_{i=1}^{n} Y_i x_i / \sum_{i=1}^{n} x_i^2$ by differentiating with respect to b.

2.

Locations	Distance in mm, X_i	Road distance in miles, Y_i
Barnstaple–Colchester	82	292
Aberdeen–Fort William	43	157
Edinburgh–Guildford	130	431
Bristol–Holyhead	58	251
Hereford–Hull	57	201
Birmingham–Inverness	133	459
Cardiff–Leeds	64	232
Carmarthen–Lincoln	71	253
Glasgow–Maidstone	144	459
Colchester–Manchester	68	212
Lincoln–Newcastle	49	157
Carmarthen–Exeter	33	183

Given the above table of towns/cities in Britain and their Y and x values as defined in Example 5, calculate the least-squares estimate of the parameter value b. Draw a scatter diagram and the fitted line through the origin. What are the units of b and how can the parameter be interpreted or used? The values $Y_i - bx_i$ are called the residuals. Why might a pair of locations such as Carmarthen–Exeter yield a large residual?

4.7 Linear regression

The topic of regression, and in particular linear regression, has been introduced in Chapter 2. In this section we introduce it by way of an experiment that you may well be able to do for yourself – all you need is a long bench with a reasonably smooth surface and a coin. Mark the bench out with lines across its width at distance 100, 150, 200, 250 and 300 cm from a starting point. Your job is to take a coin, place it at the starting line, and then 'shove it' aiming at each target distance in turn. Measure the distance the coin actually travels (to the nearest cm?) and produce a table like Table 4.4 and a scatter diagram like Fig 4.4.

Table 4.4 Shove-halfpenny data.

x	100	150	200	250	300
y	104	153	205	241	280
y	102	156	218	254	319
y	107	166	191	229	275
y	115	145	227	281	321
y	107	150	225	276	290

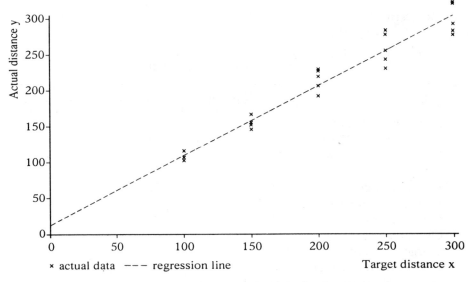

Fig 4.4 The experimental data plotted and the fitted regression line.

How do we model this experimental situation? First we need some variables – let Y_i denote the distance the coin will travel on the ith shove when it is being aimed at the target value x_i. Almost certainly you do not belong to that select band of 'shovers' who can shove to a desired point and achieve it (to the nearest cm), and so, barring freaks, Y_i never equals x_i.

Suppose, however, that you were 'right on average' for each target value. The natural interpretation of this statement using the language of statistics is that $E(Y_i) = x_i$, i.e. the expected value for each shove was equal to the target value. Now suppose this is not true. In what simple ways could this statement be false? Well, perhaps your shoves on average were always, say, 5 cm short of the target, i.e. $E(Y_i) = x_i - 5$, or more generally,

$$E(Y_i) = \beta_0 + x_i \quad \text{(model where the 'bias' is a fixed amount)}$$

Alternatively, the average bias could be a fixed proportion of the target value, e.g. $E(Y_i) = 0.95 x_i$, or more generally,

$$E(Y_i) = \beta_1 x_i \quad \text{(model where the 'bias' is a fixed percentage)}$$

These are simple, not necessarily realistic, models which are alternatives to the 'right on average model', or more correctly, they are generalisations of the model because they include the 'right on average' model as a special case ($\beta_0 = 0$ for the first generalisation and $\beta_1 = 1$ for the second generalisation).

Why should not both these cases happen? In other words, why might you not be inaccurate because your shoves, on average, equal a fixed amount plus a fixed percentage of the target, i.e. $E(Y_i) = \beta_0 + \beta_1 x_i$?

This is the linear regression model – the expected value for the shove depends on the x value through a linear relationship involving an intercept parameter β_0 and a slope parameter β_1.

EXERCISE ON 4.7

1. For each of these models, for x values in the range 100 to 300, draw on the same diagram four lines representing the situation that
   (a) you are right on average,
   (b) you are short of the target by 5 cm on average,
   (c) on average your shove reaches 95% of the target,
   (d) your expected shove equals 10 plus 90% of the target value.
   (What other models might we propose?)

The principle of least squares may be applied directly to the data in just the same way as we did before – by adopting our three-column format and constructing the sum of squares to be minimised, as in Table 4.5.

Unfortunately, the sum S now depends on two unknown parameters, so the details of its minimisation are mathematically more complicated than in earlier examples where we had only one parameter. One way to derive the least-squares estimates as given in the last row of our least-squares table above is to appeal to our intuition as follows.

Whatever the least-squares line is, it would be surprising if it did not go through the point $(\bar{x}, \bar{y})$, which means that the formula for the line must be of the form

Statistics

Table 4.5 Least-squares table for the regression models.

Observation	Expected value	Square of difference
y_1	$\beta_0 + \beta_1 x_1$	$(y_1 - \beta_0 - \beta_1 x_1)^2$
y_2	$\beta_0 + \beta_1 x_2$	$(y_2 - \beta_0 - \beta_1 x_2)^2$
$\vdots$	$\vdots$	$\vdots$
y_n	$\beta_0 + \beta_1 x_n$	$(y_n - \beta_0 - \beta_1 x_n)^2$

Sum of squares

$$S(\beta_0, \beta_1) = \sum_{i=1}^{n} (y_i - \beta_0 - \beta_1 x_i)^2$$

Least squares occurs when

$$\beta_1 = \sum_{i=1}^{n} y_i(x_i - \bar{x}) \bigg/ \sum_{i=1}^{n} (x_i - \bar{x})^2$$

and $\beta_0 = \bar{y} - \beta_1 \bar{x}$

$y = \bar{y} + \beta_1(x - \bar{x})$. This means that we can rewrite the formula for $S(\beta_0, \beta_1)$ in the form

$$S(\beta_1) = \sum_{i=1}^{n} [y_i - \bar{y} - \beta_1(x_i - \bar{x})]^2$$

which is of a very similar form to the previous example and from which we deduce, by differentiating with respect to β_1, that S is minimised when

$$\beta_1 = \sum_{i=1}^{n} (y_i - \bar{y})(x_i - \bar{x}) \bigg/ \sum_{i=1}^{n} (x_i - \bar{x})^2 = S_{xy}/S_{xx} \text{ say}$$

(where S_{xx} and S_{xy} are defined in Section 2.10).
Note that

$$S_{xy} = \sum_{i=1}^{n} (y_i - \bar{y})(x_i - \bar{x}) = \sum_{i=1}^{n} y_i(x_i - \bar{x})$$

because

$$\sum_{i=1}^{n} \bar{y}(x_i - \bar{x}) = \bar{y} \sum_{i=1}^{n} (x_i - \bar{x}) = \bar{y} \times 0 = 0$$

(This will be useful later on when we come to examine the means and expected values of the estimates.)

Finally, we need to estimate the residuals (the differences between the observed y values and the fitted y values), to square them to calculate the residual sum of squares, and (by dividing this by $n - 2$) to estimate the variance of the error term:

ith residual $= \hat{E}_i = y_i - \hat{y}_i = y_i - \bar{y} - \hat{\beta}_1(x_i - \bar{x})$

Residual sum of squares,

$$\text{RSS} = \sum_{i=1}^{n} \hat{E}_i^2 = S_{yy} - \hat{\beta}_1 S_{xy} \quad \text{so } \hat{\sigma}^2 = \frac{\text{RSS}}{n-2}$$

Example 6

Find the least-squares regression line of y on x when x takes the values 1, 2, 3, 4 and y takes corresponding values 2, 5, 5, 8.

We tackle the calculations in 'calculator mode' in three stages:

1. Preliminary calculations

$$\sum_{i=1}^{4} x_i = 10 \qquad \sum_{i=1}^{4} x_i^2 = 30 \qquad \sum_{i=1}^{4} x_i y_i = 59 \qquad \sum_{i=1}^{4} y_i = 20$$

$$\sum_{i=1}^{4} y_i^2 = 118 \qquad n = 4$$

2. Second-stage calculations

$$\bar{x} = 10/4 = 2.5 \qquad \bar{y} = 20/4 = 5$$

$$S_{xx} = \sum_{i=1}^{4} x_i^2 - \frac{1}{n}\left(\sum_{i=1}^{4} x_i\right)^2 = 30 - \frac{10^2}{4} = 5$$

$$S_{xy} = \sum_{i=1}^{4} x_i y_i - \frac{1}{n}\sum_{i=1}^{4} x_i \sum_{i=1}^{4} y_i = 59 - \frac{10 \times 20}{4} = 9$$

$$S_{yy} = 118 - \frac{20^2}{4} = 18 \quad \text{(calculated in the same way as } S_{xx}\text{)}$$

3. Final-stage calculations

$$\hat{\beta}_1 = \frac{S_{xy}}{S_{xx}} = \frac{9}{5} = 1.8 \qquad \hat{\beta}_0 = \bar{y} - \hat{\beta}_1 \bar{x} = 5 - 1.8 \times 2.5 = 0.5$$

$$\text{RSS} = 18 - 1.8 \times 9 = 1.8 \qquad \hat{\sigma}^2 = \frac{1.8}{4-2} = 0.9$$

Figure 4.5 graphs the data and the fitted line. The dotted lines indicate the residuals which are displayed in Table 4.6. Note that their sum of squares (1.8) agrees with the value of RSS calculated by different means above.

Note that the sum of the residuals is zero – this follows from the assumption used in calculating the least-squares estimates that the regression line should pass through the point $(\bar{x}, \bar{y})$.

Related to this point is the observation that the least-squares estimates have the effect of making the estimates of the error terms as nearly normally distributed as possible. Consequently it will do its best not to produce an

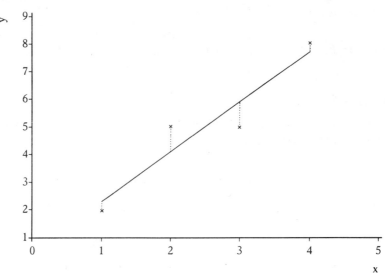

Fig 4.5 Graph of the data and fitted line for Example 6.

Table 4.6 The residuals for Example 6.

y_i	$\hat{y}_i = \hat{\beta}_0 + \hat{\beta}_1 x_i$	$\hat{E}_i = y_i - \hat{y}_i$
2	2.3	−0.3
5	4.1	0.9
5	5.9	−0.9
8	7.7	0.3

abnormally large outlier. For example, if the fourth y value is changed to 20, the equivalent table is

y_i	$\hat{y}_i = \hat{\beta}_0 + \hat{\beta}_1 x_i$	$\hat{E}_i = y_i - \hat{y}_i$
2	−0.1	2.1
5	5.3	−0.3
5	10.7	5.7
20	16.1	3.9

With an example of this nature, you would probably have come to the conclusion that the fourth value was an outlier, and largely ignored it, so that the fourth residual would have been high. Since, however, the sum of the squares of the residuals is made as small as possible (least squares!) although the sizes of the residuals have all increased (and the estimate of the variance is now 26.1) the fourth residual is not the largest! This means that in practice, a full regression analysis of data must include a graphical plot including the fitted line (a good way of checking your calculations) and an examination of the plot to determine whether particular points have undue influence on the results – if such a point is suspected, the calculation should be repeated leaving this point out to determine how the estimated regression line changes.

EXERCISE ON 4.7

2. Show that the least-squares regression line for the amended data above is given by $y = -5.5 + 5.4x$. Verify also the new table of residuals and the new estimate of the variance quoted above.

Regression calculations on a computer

In the small-scale regression analysis above, we have calculated sums of squares such as S_{xx} using the formula

$$\sum_{i=1}^{n} x_i^2 - \frac{1}{n}\left(\sum_{i=1}^{n} x_i\right)^2$$

and similarly for the other values S_{xy} and S_{yy}. These formulae should be treated with caution as they are numerically unstable. You can see the reason – the formula consists of subtracting one quantity from another, and if these are both large, as they could well be, much of the accuracy disappears in the subtraction. (Indeed hand-held calculators with a mean and standard deviation routine which unfortunately use this algorithm can compute a negative value for S_{xx}, if large enough numbers are keyed in, and hence yield an error message when the standard-deviation button is pressed!)

If on the other hand a computer is used, then the computations can proceed easily by direct calculations of the values $x_i - \bar{x}$, squaring and then summing – this is much to be preferred.

EXERCISE ON 4.7

3. For the shove-halfpenny data of Table 4.4, derive the following regression results: $\bar{x} = 200$, $\bar{y} = 205.48$, $S_{xx} = 125000$, $S_{yy} = 121904.24$, $\hat{\beta}_0 = 12.6$, $\hat{\beta}_1 = 0.9644$, $RSS = 5645.82$, $\hat{\sigma} = 15.667$.

Note: it is interesting to note that the estimate of the slope is near 1 (as we would expect it to be if our 'shoving' were accurate), but that the intercept estimate is not near 0. This difference is a delusion! We have to be careful to define 'nearness' in terms of the amount of scatter that we can expect in these estimates. This is controlled by the scatter as measured by σ (and estimated as 15.667 cm) and also by the mathematical nature of the least-squares estimates. We will look at this more closely in the next chapter, but a little intuition goes a long way! The estimate of the slope parameter should be fairly accurate because we have chosen the x values over a very wide range (in relation to σ). However, we can expect the estimate of the intercept parameter to be more variable, because there are no x measurements near 0. You should review these comments after you have read the next chapter and relate them to the formulae given there for the variances of the estimates $\hat{\beta}_1$ and $\hat{\beta}_0$.

Summary

What we have tried to do in this chapter is to present the very important mathematical basis for techniques of estimation of parameters and their associated confidence intervals. We have also tried to develop the reader's understanding of what is meant by a statistical model and its relevance to the practical analysis of data. Unbiasedness of an estimator tells us that on average it will get the right answer, and its variance has been used to rank some unbiased estimators as being more efficient than others. The principle of maximum likelihood estimation is introduced in terms of the standard statistical models presented in Chapter 3, but in addition, maximum likelihood is applied in non-standard situations where intuition might have failed. For normally distributed data, maximum likelihood estimation reduces to least-squares estimation which is presented in an elementary but methodical way so that it can be seen that estimation of population means in one- and two-sample situations is not really that different from its application to linear regression problems. It is all too easy for the subject of statistics to be seen as an incoherent muddle, when it isn't!

- An estimator of a parameter is unbiased if its sampling distribution has a population mean equal to the parameter value. Unbiased estimators with small variance are good estimators.
- Maximum likelihood estimates are derived by finding the value of the parameter that maximises the likelihood or log-likelihood function. For common distributions the estimators derived usually agree with intuitively based estimators.
- Least-squares estimates are derived by minimising the sum of squares of the difference between observations and their expected values. For normally distributed data, they are equivalent procedures.

FURTHER EXERCISES

1. Show that the least-squares regression line of yield on fertilizer for Example 10 of Chapter 2 is given by the equation $y = 1.596 + 0.926x$. Verify that the estimate of the standard deviation of each observation is $\hat{\sigma} = 1.088$. Verify also that the predicted yield for a fertilizer level of $x = 10$ is 10.851. Should you use this regression line to predict the expected yield from applying fertilizer at level $x = 20$?

2. A bean plant is measured at weekly intervals over a period of 10 weeks. Assume that the height Y_i after x_i weeks is such that $E(Y_i) = bx_i^2$. Given the data below, estimate the value of the constant b. If you assumed that each Y_i had a normal distribution with some unknown variance, could you write down a likelihood function that would give the same least-squares estimate? (**Hint:** Are the observations statistically independent?) Investigate whether you think a straight-line relationship would be adequate for the data.

x_i	1	2	3	4	5	6	7	8	9	10
y_i	1.8	2.5	6.5	7.2	8.2	9.5	12.3	17.6	18.9	24.3

3. You have an old-fashioned chemical balance consisting of two pans (L and R on either end of a beam), and a set of known weights that allow you to balance the beam. You also have two unknown weights a and b. First of all you place both weights on pan L and balance using known weights amounting to 16 g. Next you place weight a only on pan L and balance using weights totalling 11.5 g. Next you place weight b on L and balance using weights equal to 4.5 g. Finally you place a on pan L and b on pan R and balance using weights 5.5 g. Using the three-column format approach to least squares, show that the least-squares estimates of a and b are 11 g and 5 g respectively.

Hint: Derive the formula for $S(a, b)$ and minimise it by 'completing the square' twice.

TUTORIAL PROBLEM

> For this problem on maximum likelihood estimation, you will need a small polythene bag containing about 50 to 60 g of lentils (whole brown ones preferred as they are larger than red lentils and easier to handle). Remove a teaspoonful of the lentils from the bag, count them (n) and colour them somehow so that they are easily distinguishable. Now return them to the bag and mix thoroughly. Now remove a second teaspoonful of lentils, count how many there are (m) and how many were in the first sample (r). You now require to estimate the number N in the bag using n, m and r. (This is known as capture–recapture and is used by population biologists to estimate population sizes.)
>
> By reference to the hypergeometric distribution write down an expression $p(N)$ which represents the probability that r out of m in the second sample came from the first sample of size n, given N. This expression depends only on the unknown N – all other values are known to you. Hence use the principle of maximum likelihood estimation to estimate the value of N. (Note that there are different ways of approaching this problem; you could tackle it numerically by choosing arbitrary values of N and calculating $p(N)$, or you could tackle it mathematically.)

5 • Hypothesis Testing

In Chapter 4 we have discussed various methods of exploring the characteristics of a population, using samples to **estimate** parameters. We used **point estimates** such as $\bar{x}$ as our 'best guess' at, for example, the value of the population mean μ, and we used **interval estimates** such as 95% confidence intervals for μ. Our general approach has been to 'ask the data' what sort of value we should adopt for a parameter.

In practice we often collect data with the express purpose of proving or disproving a theory or claim, which may itself involve specifying values, or ranges of values, for parameters. The question we then ask the data is not 'What value should we adopt?' but rather 'Is the value suggested by our theory acceptable?'

In this chapter we shall look at the ways in which we can express theories, or **statistical hypotheses**, concerning values of parameters, and the **hypothesis tests** that statisticians have devised to attempt to support or discredit them.

We shall approach the problem in three ways. The first is a straightforward application of confidence intervals. The second is the classical formulation in which a test statistic is calculated and tested at a fixed probability (**significance level**). The third, which we shall then develop for a variety of applications, involves calculating a **significance probability** for the test statistic.

5.1 The confidence interval approach

One way we can test a claim concerning the value of a parameter is to take a random sample and calculate a confidence interval (say 95%) for the parameter in question. If the proposed value falls inside the confidence interval we can accept it, otherwise we can reject it.

Example I

The manufacturers of pre-packaged sweets sell a product labelled 500 g. They claim that their product has a mean weight of 520 g and a standard deviation of 10 g. A consumers' association weights and measures department tests 100 randomly selected units and finds a mean of 516.2 g. Is this result consistent with the manufacturers' claim?

Call the true (population) mean μ and the sample mean $\bar{x}$. The manufacturers' claim is that $\mu = 520$. It may or may not be reasonable to assume that the weights are normally distributed, but the central limit theorem tells us that the **mean** of a large number of values should be approximately normally distributed. Using the sample mean and the manufacturers' claimed standard deviation, a 95% confidence interval for μ is $516.2 \pm 1.96 \times 10/\sqrt{100} = 516.2 \pm 1.96$, i.e. (514.24, 518.16). This interval does not include the value claimed by the manufacturers, so we have some grounds for rejecting their claim.

Of course, the limits of the confidence interval we have calculated are themselves random variables because of the random sampling carried out. If we repeated the experiment, and measured another 100 units, the new confidence interval, based on a new sample mean, may include 520. But we do expect 19 out of 20 of such intervals to contain the *true* value of μ, so our result represents quite strong evidence that 520 is *not* the true value.

Another question to consider is – why a 95% confidence interval; why not 90% or 99%? Let us first calculate a 99% confidence interval for Example 1. The point below which 99.5% of an $N(0, 1)$ distribution lies is 2.576, so the interval is $516.2 \pm 2.576 \times 10/\sqrt{100} = 516.2 \pm 2.576$, i.e. (513.62, 518.78). The result is a larger interval than before, but still not large enough to include 520. As we would expect 99% of such intervals to include the true value of μ, we now have evidence suggesting that 520 is *not* the correct value. If we make the confidence level sufficiently high, eventually the claimed value $\mu = 520$ would be included in the confidence interval. So what level of confidence should we choose? The 95% level is arbitrary but very much an accepted convention, or default choice. The user of confidence intervals is, however, free to choose a percentage to reflect the relative importance attached to (i) accepting a claim that may really be false and (ii) rejecting a claim that may really be true, but this is beyond the scope of this book.

EXERCISE ON 5.1

1. The time taken by statistics students to complete a computer practical session is normally distributed with mean 37.5 minutes and standard deviation 5.5 minutes. I test a particular group of 10 students who have never used a computer before and find that they take an average of 42.3 minutes to complete the session. On the basis of this evidence, should I provide extra tuition to these students in the use of the computer?

5.2 Classical formulation of hypothesis tests

A statistical hypothesis is a statement concerning the distribution of a random variable, or the probabilities involved in a statistical experiment. In Example 1, the manufacturers' claim that the mean is 520 g is a hypothesis concerning the distribution of the weight of the packets of sweets. Using the central limit theorem we claim that $\bar{X}$ has a normal distribution with mean 520 and standard deviation 1. This sort of hypothesis, which completely specifies the probabilities, is called a **simple hypothesis**.

The hypothesis that the mean is *at least* 520 does not completely specify the parameters, and is called a **composite hypothesis**.

The null and alternative hypotheses

The purpose of the test is to choose between two hypotheses. One is called the **null** hypothesis (denoted H_0) and the other is the **alternative** hypothesis (denoted

H_1). Generally, the null hypothesis is the one we are attempting to *disprove*. We write

H_0: $\mu = 520$ (null hypothesis)
H_1: $\mu \neq 520$ (alternative hypothesis)

The decision rule

We must now formulate a rule for deciding between the two hypotheses. This **decision rule** is a function of the data. A typical decision rule is 'accept H_0 (that the manufacturers' claim is true) if $\bar{x}$ is between 518 and 522', which makes sense in that the further $\bar{x}$ is away from 520, the more we would regard the evidence to be against H_0.

The critical region

The set of values of the test statistic which cause us to reject H_0 is called the **critical region**. For our decision rule, the critical region includes values less than 518 and values greater than 522 (see Fig 5.1).

Two types of error and their probabilities

It is possible for $\bar{x}$ to be less than 518 or greater than 522 when the true mean is 520 and, similarly, it is possible for $\bar{x}$ to be in the interval 518 to 522 when the true mean is 515, so the above rule could lead to an incorrect conclusion in two ways:

- The incorrect rejection of H_0 (saying μ is not 520 when it really is) is called a **Type I error**.
- The incorrect acceptance of H_0 (saying μ is 520 when it really is not) is called a **Type II error**.

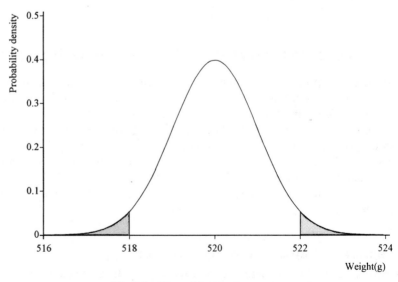

Fig 5.1 The critical region.

Hypothesis Testing

The probabilities associated with Type I and Type II errors are determined by the choice of critical region, and are generally denoted by the Greek letters α and β respectively. The probability α of a Type I error is also called the **significance level** or **size** of the hypothesis test. The value of β depends on a specific alternative hypothesis, e.g. that $\mu = 510$ g, and is called the **power** of the test. We want to keep α and β as small as possible, but any decrease in α leads to an increase in β, unless we increase the sample size (in this case the number of packets tested). We usually select a value such as 0.05 or 0.01 for α, the significance level, and determine the critical region accordingly.

What is α for our decision rule? If H_0 is true,

$$P(\bar{X} < 518) = \Phi\left(\frac{518 - 520}{10/\sqrt{100}}\right) = \Phi(-2) = 0.02275$$

$$P(\bar{X} > 522) = 1 - \Phi\left(\frac{522 - 520}{10/\sqrt{100}}\right) = 1 - \Phi(2) = 0.02275$$

So $\alpha = 0.02275 + 0.02275 = 0.0455$.

If $\bar{x} < 518$ or $\bar{x} > 522$ we can therefore reject the hypothesis that $\mu = 520$ at the 4.55% significance level, i.e. the probability of getting such an extreme result if H_0 is true is 0.0455.

Suppose we want to formulate a decision rule with a significance level of 5%. Under H_0,

$$\bar{X} \sim N(520, 1)$$

so

$$Z = \frac{\bar{X} - 520}{10/\sqrt{100}} \sim N(0, 1)$$

Given an observed value $\bar{x}$ for random variable $\bar{X}$, we can therefore calculate an observed value z for random variable Z:

$$z = \frac{\bar{x} - 520}{10/\sqrt{100}}$$

Recall that $\Phi(1.96) = 0.975$, so $P(Z < -1.96 \text{ or } Z > 1.96) = 0.05$. So a decision rule with significance level 5% is

Reject H_0 if $z < -1.96$ or $z > 1.96$

Now

$$z > 1.96 \text{ if } \frac{\bar{x} - 520}{10/\sqrt{100}} > 1.96 \text{ (i.e. if } \bar{x} > 521.96)$$

and

$$z < -1.96 \text{ if } \frac{\bar{x} - 520}{10/\sqrt{1000}} < -1.96$$

(i.e. if $\bar{x} < 518.04$). So another form of our decision rule is

Reject H_0 if $\bar{x} < 518.04$ or if $\bar{x} > 521.96$

With our observed value, $\bar{x} = 516.2$, we therefore reject the manufacturers' claim at the 5% significance level. Note that this form of test will always give the *same conclusion* as the equivalent 95% confidence interval form of test.

More generally, we reject $H_0: \mu = \mu_0$ in favour of $H_1: \mu \neq \mu_0$ if

$$\bar{x} < \mu_0 - 1.96 \frac{\sigma}{\sqrt{n}}$$

or

$$\bar{x} > \mu_0 + 1.96 \frac{\sigma}{\sqrt{n}}$$

Example 2

Sales of a particular make of washing machine at an electrical appliance store produce a weekly revenue with a mean of £4610 and standard deviation £140. The manufacturing firm is taken over by another organisation and the sales during the following six weeks average £4770. Test at the 1% significance level whether the take-over has affected sales, assuming that the revenue is normally distributed and that the standard deviation is still £140.

We shall adopt the null hypothesis H_0 that there is no change in the distribution of the revenue, i.e. we expect the mean $\bar{X}$ of the six weeks after the take-over to be $N(4610, 140^2/6)$.

Define

$$Z = \frac{\bar{X} - 4610}{140/\sqrt{6}}$$

Under H_0, $Z \sim N(0, 1)$. Our calculated value of Z is

$$\frac{4770 - 4610}{140/\sqrt{6}} = 2.799$$

This exceeds the critical point 2.58, below which 99.5% of $N(0, 1)$ values should lie, so we reject H_0 at the 1% significance level. In other words, the six weeks of sales figures since the take-over provide sufficient statistical evidence for us to conclude that the mean revenue has *increased* (since the new estimated mean is higher than the old one).

One-tailed and two-tailed hypothesis tests

Note that in Example 2 we tested for a *change* in the revenue, and we were therefore prepared to find an increase or a decrease as a significant result. This sort of test, where rejection of the null hypothesis can take place at either end of the scale, is called a **two-tailed test**.

If it is reasonable for us to believe a priori that the take-over is *expected* to lead to an *increase* in revenue, we could test specifically for an increase, by defining our critical region on one side only. For a 1% significance level this involves using a critical value for Z of 2.33, below which 99% of $N(0, 1)$ values may be expected to lie. For a 95% significance level, the corresponding critical value is 1.64. Tests based on these critical values are called **one-tailed tests**.

Using a one-tailed test for Example 2, we need a Z value that exceeds 2.33 (which of course our value of 2.799 does). Note, however, that a Z value of -2.799 would *not* be counted significant in such a test, as we are now specifically testing for an increase in revenue.

It is very important to note that the *reason* for choosing a one-tailed test must be independent of the data collected, and ideally we should make such a choice before collecting, or at least before looking at, the data. The fact that the new value of $\bar{x}$ is larger than the previous mean must *not* contribute to the decision to use a one-tailed test.

5.3 Hypothesis tests for discrete data

Suppose that a medical researcher is sceptical about the effectiveness of the treatment applied to the patients in the cholesterol experiment (Dataset 1, Appendix A). The researcher may put forward the hypothesis that each patient is just as likely to increase cholesterol level as to decrease. This is equivalent to the statement 'the probability, p, that *chola* < *cholb* is 0.5'. We shall take this as our null hypothesis, with the alternative that the treatment 'has some effect':

H_0: $p = 0.5$ (null hypothesis)
H_1: $p \neq 0.5$ (alternative hypothesis)

Let us define a 'success' to be a patient who improves (*chola* < *cholb*) and a 'failure' a patient who does not improve (*chola* $\geq$ *cholb*). We shall also take the complete dataset of 118 values as our 'experiment'.

Suppose we believe that the treatment will have some effect (either beneficial or harmful). One way to test whether we are correct is to apply the treatment to a large number of patients and count the number of successes. If approximately half the patients improve, we must admit that the treatment probably has no effect, but if a lot more than half improve (or vice versa) we are probably safe if we conclude that it has had some effect.

A possible decision rule would be 'accept H_0 (that the treatment has no effect) if we obtain between 50 and 70 successes'. So our critical region includes 0–49 successes and 71–118 successes.

What is α for our decision rule? Let X be the number of successes. If H_0 is true, X will have a binomial distribution with parameters $n = 118$, $p = 0.5$. Let us also define random variables Y, normally distributed with mean np and variance $np(1-p)$, i.e. $Y \sim N(59, 29.5)$, and $Z \sim N(0,1)$.

Now $\alpha = P(X < 50 \text{ or } X > 70)$.

Using the normal approximation to the binomial,

$$\alpha = P(Y < 49.5 \quad \text{or} \quad Y > 70.5)$$
$$= P\left(Z < \frac{49.5 - 59}{\sqrt{29.5}} \quad \text{or} \quad Z > \frac{70.5 - 59}{\sqrt{29.5}}\right) = \Phi(-1.749) + 1 - \Phi(2.117)$$
$$= 0.057$$

If $X < 50$ or $X > 70$ we can therefore reject the hypothesis that the treatment has no effect at the 5.7% significance level, i.e. the probability of getting such an extreme result if H_0 is true is 0.057.

Suppose we want to formulate a decision rule with a significance level of 5%. From the outset it should be noted that meaningful rules concerning X can only involve integral values (i.e. whole numbers of patients) so precise target values of significance levels may be impossible to achieve by the use of a single critical region.

Recall that $P(Z < -1.96 \text{ or } Z > 1.96) = 0.05$.

We can translate this into a symmetrical acceptance region for an $N(59, 29.5)$ random variable of the form $P(Y < r \text{ or } Y > s) = 0.05$, i.e.

$$\frac{s - 59}{\sqrt{29.5}} = 1.96 \quad \text{and} \quad \frac{r - 59}{\sqrt{29.5}} = -1.96$$

So $s = (5.431 \times 1.96) + 59 = 69.6$ and $r = (5.431 \times -1.96) + 59 = 48.4$.

In other words, $P(Y < 48.4 \text{ or } Y > 69.6) = 0.05$. This suggests that to achieve a significance level of approximately 5%, we need a decision rule which accepts H_0 if X is between 48 and 70. The true significance level of this decision rule is

$$P\left(Z < \frac{47.5 - 59}{\sqrt{29.5}} \text{ or } Z > \frac{70.5 - 59}{\sqrt{29.5}}\right) = \Phi(-2.117) + 1 - \Phi(2.117) = 0.034$$

It is therefore a 'conservative' test – with significance level ≤ 0.05.

To implement this test we calculate our **test statistic**. In this case, this could simply be X, the number of cases where *chola* < *cholb*. The answer is 87, which is well into the critical region. So there is a great deal of evidence that the treatment has an effect, and the fact that X exceeds the predicted range (i.e. there are too many cases where the cholesterol level has been reduced) allows us to state that the treatment is beneficial.

Alternatively, we could calculate the z-statistic

$$Z = \frac{X - 59}{\sqrt{29.5}}$$

Under H_0, Z would be expected to have an $N(0, 1)$ distribution. For $X = 87$, Z has the value 5.155. If we compare this with the 'acceptance' interval $(-1.96, 1.96)$ we see that it is well outside. Our conclusions are exactly the same as above. Traditionally H_0 is rejected if the *absolute value* of Z exceeds 1.96 (the **critical value**, the 0.975 quantile of the standard normal distribution). A property of this test statistic is that the test is automatically conservative.

5.4 Calculating the significance probability

Returning to Example 2 (with a two-tailed test) rather than fixing the significance level at 1% or some other value, we can try to discover the smallest significance level at which H_0 would be rejected. Using statistical tables, we can achieve this by looking up the value of the z-statistic

$$z = \frac{4770 - 4610}{140/\sqrt{6}} = 2.799$$

in the normal distribution table and reading off the corresponding probability, which is 0.9974. This is $P(Z < 2.799)$. So $P(Z > 2.799) = 1 - 0.9974 = 0.0026$ and $P(Z < -2.799) = 0.0026$. Hence $P(|Z| < 2.799) = 1 - 2 \times 0.0026 = 0.9948$.

The smallest significance level at which our test would reject H_0 is therefore equal to $1 - 0.9948 = 0.0052$, or 0.52%.

We call this value the **significance probability** of the hypothesis test. It is the probability of obtaining the *value actually observed, or a more extreme value*, if H_0 is true. If the significance probability is smaller than α, a fixed level test of size α would reject H_0. So the significance probability approach to hypothesis testing embodies all the information provided by fixed level testing at *any* significance level.

The value 0.9974 is of course simply $\Phi(2.799)$ and can be obtained from computer packages such as Minitab (in which the *default* distribution is the normal distribution):

```
MTB > cdf 2.799
       2.7990      0.9974
```

One-tailed significance probability

If we decide that a one-tailed test is appropriate, the significance probability is simply $P(Z > 2.799) = 0.0026$. For a symmetrical pdf such as the normal distribution, the one-tailed significance probability is always half of the two-tailed significance probability.

5.5 The normal distribution with unknown variance

Let us return to Example 1, and consider again the claim made by the manufacturers. As well as claiming that the mean is 520 g, they have stated that the standard deviation is 10 g. Suppose the standard deviation is really 20 g – how would that affect our calculations?

Our new confidence interval would be $516.2 \pm 1.96 \times 20/\sqrt{100} = 516.2 \pm 3.92$, i.e. (512.28, 520.12). This includes 520 (just!) and so the claim would not be rejected. So our investigations do not necessarily disprove the claim that $\mu = 520$, but they do show that *some part* of the manufacturers' claim is probably incorrect.

One way for us to proceed is to calculate the *sample* standard deviation, s, and see if it is about 10. If not, perhaps s could be used in our confidence interval calculation.

The confidence intervals we have calculated for μ depend upon the fact that

$$\frac{\bar{X} - \mu}{\sigma/\sqrt{n}} \sim N(0, 1)$$

If σ is unknown, an obvious alternative is to replace it by the sample standard deviation, s. So we consider the statistic

$$t = \frac{\bar{X} - \mu}{s/\sqrt{n}}$$

Student's t distribution

If n is large, s gives a very reliable estimate of σ and so the t-statistic will have a distribution very close to a normal distribution. For smaller values of n, when we replace σ by s we introduce extra variation. The result is a symmetrical distribution which looks very similar to the normal distribution, but is actually broader, with more of the distribution in the tails. The exact shape of the distribution depends on n.

Degrees of freedom

We call this distribution a Student's t distribution with $n - 1$ degrees of freedom. We are not concerned here with the mathematical derivation of statistical distributions, but it is important to grasp the idea of degrees of freedom. Figure 5.2 illustrates the probability density functions of t for $n = 4$ and $n = 20$ (3 and 19 degrees of freedom respectively) and contrasts them with the standard normal distribution.

Although the differences between the curves are quite small, it is important that we identify the right curve for our particular statistic. In general, to obtain the degrees of freedom, we subtract the *number of parameters estimated* (not including the standard deviation) from the total number of observations. In this case we have n observations and we have to estimate just one parameter, so there are $n - 1$ degrees of freedom.

In statistical tables, percentage points for the t distribution are usually tabulated for various numbers of degrees of freedom, and for a small number of fixed significance levels (e.g. 5% and 1%). They therefore allow us to construct confidence intervals and perform fixed level hypothesis tests, but we cannot use them to determine significance probabilities. Statistical computer packages are more flexible and will calculate any percentiles, allowing us to construct *any*

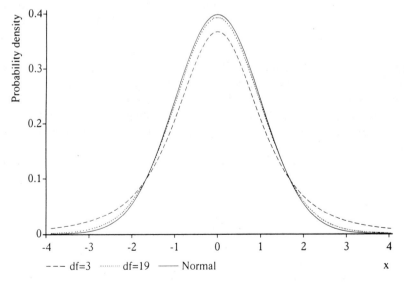

Fig 5.2 Student's t distribution and the normal distribution.

Hypothesis Testing

confidence intervals, and also the cdf at any point, from which we calculate the significance probability.

Confidence intervals for the t distribution

To calculate a 95% confidence interval for μ, note that $P(-t_{0.975} < t < t_{0.975}) = 0.95$ (where $t_{0.975}$ is the 97.5th percentile of the t distribution with $n-1$ degrees of freedom). So a 95% confidence interval for μ is $\bar{X} \pm t_{0.975} s/\sqrt{n}$.

Example 3

A beer-dispensing pump is supposed to deliver 1 pint (20 fluid ounces) of beer. The amount actually dispensed is thought to be normally distributed. Ten 'pints' are measured accurately, with the following results:

19.96 19.97 19.94 20.01 19.99 19.97 19.95 19.97 20.00 19.98

Construct a 95% confidence interval for the mean μ.

$$\bar{x} = \frac{1}{10}\sum_{i=1}^{10} x_i = 19.974 \quad \text{and} \quad s = \sqrt{\frac{1}{9}\sum_{i=1}^{10}(x_i - \bar{x})^2} = 0.0217$$

The 97.5th percentile of the t distribution with $n-1 = 9$ degrees of freedom is 2.262. So a 95% confidence interval for μ is

$$\bar{x} \pm t_{0.975} s/\sqrt{n} = 19.974 \pm 2.262 \times 0.0217/\sqrt{10}$$
$$= 19.974 \pm 0.0155 = (19.96, 19.99)$$

This interval does not include the value 20. Although all the values are very close to 20, the small spread of the values makes a true mean of 20 most unlikely.

Student's t-test

In Example 3 we can formulate a null hypothesis $H_0: \mu = 20$ and an alternative $H_1: \mu \neq 20$.

Under H_0, the statistic

$$t = \frac{\bar{X} - \mu}{s/\sqrt{n}}$$

would have a t distribution with $n-1$ ($= 9$) degrees of freedom. Values of t in the interval $(-t_{0.975}, t_{0.975})$ would allow us to accept H_0 at the 5% significance level (i.e. $|t| < t_{0.975}$). But values such that $|t| > t_{0.975}$ cause us to reject H_0.

For our data,

$$t = \frac{19.974 - 20}{0.0217/\sqrt{10}} = \frac{-0.026}{0.00686} = -3.79$$

and from tables $t_{0.975} = 2.262$. So $|t| > t_{0.975}$ and we reject the null hypothesis that the true mean is 20.

From Minitab, we can find the tail probability:

```
MTB > cdf -3.79;
SUBC> t 9.
    -3.7900      0.0022
```

So for a two-tailed test the significance probability is $2 \times 0.0022 = 0.0044$.

EXERCISES ON 5.5

1. A garage wants its petrol price to be lower than the average in its area. Its current price is 55p per litre. The garage takes a random sample of 10 garages in the area and obtains the following prices:

    59 53 57 60 55 50 53 57 57 55

    Test at 5% significance level whether the average (of all garages in the area) is higher than this garage. Construct a 95% confidence interval for the mean price.
2. On entry into a school, students are grouped according to the results of an IQ test, the brightest in class A, average students in class B, below average in class C. The average IQs are 125, 105 and 95, respectively. A random sample of five students for one class is tested and found to have the following IQs:

    106 98 101 97 100

    Construct a 95% confidence interval for the mean μ of the class for which they were selected. Can you identify the class with 95% confidence?

5.6 Matched pairs

We often want to test whether a particular treatment has a specific effect. Examples are:

- Does regular exercise reduce an employee's amount of leave due to illness?
- Do revision and tutorial classes increase the marks obtained by a student in examinations?
- Does an increase in bus fares reduce the frequency of trips people make into town?

We can investigate all of these by taking measurements on the same set of subjects before and after the 'treatment':

- Count the number of days of illness for each employee in one year; institute the regular exercise programme; and then count the number of days of illness in the next year.
- Give the students one examination; administer the tutorials and revision programme; and then give another examination.
- Count the number of trips per person (for a month, say); increase the fares; and count the number of trips for another month.

In each case we have two observations per subject, and we want to know whether the expected effect is observed. There may be considerable variation between subjects, e.g. some people may go to town once a fortnight, others may go five times a week. To take account of this variation, for each subject we subtract the 'before treatment' measurement from the 'after treatment' measurement to obtain a single set of data – the **differences**. We then apply a hypothesis test (such as a *t*-test) to the differences, with the null hypothesis that there is no significant change (i.e. the mean difference is not significantly different from zero) – $H_0: \mu = 0$.

Example 4

Ten students fail their examinations and are required to resit. After a period of revision and tutorials they take a fresh examination. Their marks are listed in Table 5.1.

Table 5.1 Student examination results.

Student number (i)	Examination (X_i)	Resit (Y_i)	Difference ($D_i = Y_i - X_i$)
1	30	40	10
2	32	34	2
3	27	50	23
4	30	38	8
5	35	33	−2
6	32	45	13
7	25	40	15
8	20	30	10
9	31	38	7
10	30	42	12

Were the revision and tutorials effective?

Here we need a one-tailed test, since we are expecting an improvement in the marks. In Minitab, putting X and Y into c1 and c2 respectively:

```
MTB > let c3=c2-c1
MTB > ttest c3

TEST OF MU = 0.000 VS MU N.E. 0.000

            N       MEAN     STDEV    SE MEAN        T    P VALUE
C3         10      9.800     6.893      2.180     4.50    0.0015
```

We have observed an improvement (the mean difference is 9.8 marks) and the (two-tailed) p value of 0.0015 shows that this is highly significant (not the sort of improvement that could be achieved 'by chance'). For a one-tailed test, $p = 0.00075$.

EXERCISES ON 5.6

1. A publishing firm sends 10 of its employees to a speed-reading course. Their scores (words per minute) before and after the course are as follows:

Employee	1	2	3	4	5	6	7	8	9	10
Before	950	990	1011	985	992	1001	947	962	870	968
After	1041	1030	1020	1015	995	999	1052	980	950	970

Is the course effective? Construct a 95% confidence interval for the mean improvement.

2. In comparing two methods of chlorinating sewage, eight pairs of batches of sewage were treated. Each pair was taken on a different day, the two batches on any one day being taken close together in time, and the two treatments were randomly assigned to the two batches in each pair. Treatment A involved an initial period of rapid mixing, while treatment B did not. The results, in log coliform density per ml, were as follows:

Day 1	Day 2	Day 3	Day 4	Day 5	Day 6	Day 7	Day 8
A 2.8	B 3.1	B 3.4	A 3.6	B 2.7	B 2.9	B 3.5	A 2.6
B 3.2	A 3.1	A 2.9	B 3.5	A 2.4	A 3.0	A 3.2	B 2.8

Use a paired t-test to determine whether one treatment is significantly better than the other.

3. Using the cholesterol data (Dataset 1, Appendix A), use a paired t-test to test whether there is a significant reduction in (a) cholesterol level ($cholb - chola$), and (b) weight ($wtb - wta$).

5.7 Differences of means (two independent samples)

In Section 5.6 we considered three situations and designed 'matched-pairs' trials to test for the effects which were of interest. In many practical situations this approach is either impossible or undesirable. For example, we may want to test whether a drug prolongs survival time. We cannot observe survival time without the drug and then administer the drug and observe the new survival time, since we have to allow the subject to die in order to observe the survival time. If we want to investigate an effect of smoking, it is not socially acceptable to persuade non-smokers to become smokers for the second half of the matched-pairs experiment!

In such cases, we must look at two independent groups of subjects (samples), generally chosen to be as similar as possible. Our two samples do not necessarily have to be the same size. Two-sample solutions to the three matched-pairs problems are:

- Divide the employees randomly into two groups; give one group the exercise programme; and count the number of days of illness for each employee.

- Divide the students randomly into two groups; give one group tutorials and revision; and then give all the students the same examination.
- Select two groups of people, one group from each of two districts with a bus service into the town; increase the fares from one district only; and compare the number of trips per person in each district.

If two samples $X_1, X_2, \ldots, X_m$ and $Y_1, Y_2, \ldots, Y_n$ are independently normally distributed, with the same variance (a known value, σ^2), we can calculate the distribution of the statistic $\bar{X} - \bar{Y}$ (the difference between their means): $X_i \sim N(\mu_x, \sigma^2)$ and $Y_i \sim N(\mu_y, \sigma^2)$ so $\bar{X} \sim N(\mu_x, \sigma^2/m)$ and $\bar{Y} \sim N(\mu_y, \sigma^2/n)$. Therefore

$$\bar{X} - \bar{Y} \sim N\left(\mu_x - \mu_y, \sigma^2\left[\frac{1}{m} + \frac{1}{n}\right]\right) \text{ or } \frac{\bar{X} - \bar{Y} - (\mu_x - \mu_y)}{\sigma\sqrt{1/m + 1/n}} \sim N(0, 1)$$

We may want to test the null hypothesis that there is really no difference between the means of the two samples, $H_0: \mu_x = \mu_y$. Under this hypothesis, the μ_x and μ_y cancel each other:

$$\bar{X} - \bar{Y} \sim N\left(0, \sigma^2\left[\frac{1}{m} + \frac{1}{n}\right]\right) \text{ or } Z = \frac{\bar{X} - \bar{Y}}{\sigma\sqrt{1/m + 1/n}} \sim N(0, 1)$$

To test $H_0: \mu_x = \mu_y$ at the 5% significance level against one-tailed or two-tailed alternative hypotheses, evaluate Z and reject H_0 if:

(1) $|Z| > 1.96$ (two-tailed);
(2) $Z > 1.64$ (one-tailed, positive critical region);
(3) $Z < -1.64$ (one-tailed, negative critical region).

A symmetrical 95% confidence interval for the difference of means $\mu_x - \mu_y$ is

$$\bar{X} - \bar{Y} \pm 1.96\sigma\sqrt{\frac{1}{m} + \frac{1}{n}}$$

Example 5

Using the data from the cholesterol experiment (Dataset 1, Appendix A), assume that the weights of patients before the treatment have a standard deviation of 14 kg, and test whether male patients are significantly heavier than female patients.

Call the *wtb* value for male i X_i and the *wtb* value for female j Y_j. There are 58 males and 60 females, so $m = 58$ and $n = 60$. All we need from the data are the two means $\bar{X}$ and $\bar{Y}$, which are 80.80 kg and 71.23 kg, respectively. Then

$$Z = \frac{\bar{X} - \bar{Y}}{\sigma\sqrt{1/m + 1/n}} = \frac{80.80 - 71.23}{14\sqrt{1/58 + 1/60}} = 3.71$$

From tables, or using the computer, a z value of 3.71 is highly significant, so clearly the male patients are significantly heavier than the female patients.

Difference of means with unknown variance

The main problem with all of these calculations is that in practice we do not usually know the variance. Suppose our two samples $X_1, X_2, \ldots, X_m$ and $Y_1, Y_2, \ldots, Y_n$ are independently normally distributed with the same variance, σ^2, but we do not know the value of σ^2. As before,

$$\frac{\bar{X} - \bar{Y}}{\sigma\sqrt{1/m + 1/n}} \sim N(0, 1)$$

but we now need to replace σ by an **estimate**, s. There are two statistics available to estimate σ^2:

$$s_x^2 = \frac{1}{m-1}\sum_{i=1}^{m}(X_i - \bar{X})^2 \quad \text{and} \quad s_y^2 = \frac{1}{n-1}\sum_{i=1}^{n}(Y_i - \bar{Y})^2$$

For mathematical reasons, however, if σ^2 is assumed common to both populations, we need to combine the information in both samples about σ^2. We do this by defining

$$s^2 = \frac{(m-1)s_x^2 + (n-1)s_y^2}{m+n-2} \quad \text{or} \quad s^2 = \frac{\sum_{i=1}^{m}(x_i - \bar{x})^2 + \sum_{j=1}^{n}(y_i - \bar{y})^2}{m+n-2}$$

For example, if $x_1 = 1$, $x_2 = 2$, $x_3 = 3$ and $y_1 = 1$, $y_2 = 3$, $y_3 = 3$, $y_4 = 5$, then $\bar{x} = 2$, $\bar{y} = 3$, $s_y^2 = 8/3$, and so

$$s^2 = \frac{2 \times 1 + 3 \times 8/3}{5} = 2$$

or, more directly,

$$s^2 = \frac{\overbrace{(1-2)^2 + (2-2)^2 + (3-2)^2}^{x \text{ terms}} + \overbrace{(1-3)^2 + (3-3)^2 + (3-3)^2 + (5-3)^2}^{y \text{ terms}}}{3 + 4 - 2}$$

$$= 2$$

Notice that in the numerator of the expression s^2 above, the means subtracted are $\bar{x} = 2$ for the x data and $\bar{y} = 3$ for the y data. This is the 'within-groups sum of squares' mentioned in Section 4.6. It has $m + n - 2$ degrees of freedom because there are $m + n$ observations and two parameters (the two means) estimated. So we can write s^2 as the quotient

$$\frac{\text{Within-groups sum of squares}}{\text{Degrees of freedom}}$$

This formulation has many more general applications.

We can replace σ by the estimator s, to obtain the statistic

$$t = \frac{\bar{X} - \bar{Y}}{s\sqrt{1/m + 1/n}}$$

which has a t distribution with $m + n - 2$ degrees of freedom if our hypothesis $H_0: \mu_x = \mu_y$ is true. A 95% confidence interval for the difference $\mu_x - \mu_y$ is therefore

$$\bar{X} - \bar{Y} \pm t_{0.975}s\sqrt{\frac{1}{m} + \frac{1}{n}}$$

where $t_{0.975}$ is the 97.5th percentile of a t distribution with $m+n-2$ degrees of freedom.

Example 6

The petrol additive *Superfuel* is supposed to reduce the petrol consumption of cars. A company runs a fleet of 30 identical cars. For a period of two months, 20 of the cars are run on petrol with *Superfuel*, and the other 10 on ordinary petrol. At the end of the trial, the petrol consumption is calculated for each car by determining the average number of miles travelled per gallon of petrol consumed (mpg). The cars on ordinary petrol give an average mpg of 38.2, with standard deviation 5.3, while the *Superfuel* cars average 45.6, with standard deviation 4.7. Calculate a symmetrical 95% confidence interval for the difference between the two means, and test the hypothesis that *Superfuel* affects the petrol consumption of cars.

Call the *Superfuel* mileages x_i ($i = 1, 2, \ldots, 20$) and the others y_j ($j = 1, 2, \ldots, 10$). Then we are given the following statistics: $\bar{x} = 45.6$, $\bar{y} = 38.2$, $s_x = 4.7$, $s_y = 5.3$ and, of course, $m = 20$ and $n = 10$. So

$$s^2 = \frac{(m-1)s_x^2 + (n-1)s_y^2}{m+n-2} = \frac{19 \times 22.09 + 9 \times 28.09}{28} = 24.02$$

and

$$s = \sqrt{24.02} = 4.90.$$

From statistical tables, $t_{0.975}(28) = 2.048$.

A 95% confidence interval for the difference between the means, $\mu_x - \mu_y$, is given by

$$\bar{X} - \bar{Y} \pm t_{0.975}s\sqrt{\frac{1}{m}+\frac{1}{n}} = 45.6 - 38.2 \pm 2.048 \times 4.90\sqrt{\left(\frac{1}{20}+\frac{1}{10}\right)}$$

$$= 7.4 \pm 3.887 = (3.513, 11.287)$$

In other words, we have reason to believe that cars using *Superfuel* will reduce their fuel consumption, and obtain between 3.5 and 11.3 more miles per gallon. This is a test at the 5% significance level of the hypothesis that *Superfuel* affects the petrol consumption of cars, and since the interval does not contain the value 0, we conclude that there *is* an effect. We can, however, devise a more appropriate test. Since we are told that *Superfuel* is supposed to *reduce* petrol consumption, we can reasonably use a one-tailed test. We start by calculating

$$t = \frac{\bar{X} - \bar{Y}}{s\sqrt{1/m + 1/n}} = \frac{45.6 - 38.2}{4.9\sqrt{1/20 + 1/10}} = 3.899$$

Under the null hypothesis, $H_0: \mu_x = \mu_y$, we would expect this statistic to have a t distribution with $m+n-2 = 28$ degrees of freedom.

For a one-tailed test at (say) the 1% significance level, we now look up $t_{0.99}$ with 28 degrees of freedom and find a value of 2.467. So any value greater than 2.467 will be significant at the 1% significance level. Note that we use $t_{0.99}$ not $t_{0.995}$ which would give us a two-tailed test.

Alternatively, we can use a computer to calculate the significance probability, e.g. in Minitab:

```
MTB > cdf 3.899;
SUBC> t 28.
       3.8990      0.9997
```

The area to the *left* of 3.899 in the appropriate t distribution is 0.9997, so the area to the *right* of 3.899 (the significance probability) is $1 - 0.9997 = 0.0003$. Assuming that the test was conducted fairly (i.e. the cars were allocated randomly to the two groups, and the groups were treated in a similar manner), we can therefore be very confident indeed that *Superfuel* is effective.

EXERCISES ON 5.7

1. A car hire firm is trying to decide which kind of tyre to use. It has narrowed the choice down to two types, A and B. A recent study of the durability of the two types examined 15 tyres of each type, testing to destruction on a machine. The data gives the number of hours on the machine to failure. Which type would you recommend?

Tyre A	3.82	3.11	4.21	2.64	4.16
	3.91	2.44	4.52	2.84	3.26
	3.74	3.04	2.56	2.58	3.15
Tyre B	4.16	3.92	3.94	4.22	4.15
	3.62	4.11	3.45	3.65	3.82
	4.55	3.82	3.85	3.62	4.88

   On the basis of this data construct a 95% confidence interval for the mean difference in hours to failure. If you wished to halve the width of the confidence interval whilst still maintaining the confidence level at 95% how many more samples would you need to take of each type?

2. If you were the managing director of the car hire firm in Exercise 1 above and could afford to purchase 20 tyres to test under usual operating conditions, in order to choose between two types of tyre, list some possible ways of proceeding. Produce a small report to suggest what you think is the best way of obtaining as much information as possible.

3. A computer centre manager would like to know whether there is a difference in the weekly computer time (in seconds) used by employees in the finance department, and those in the legal aid department. With the following data taken from 80 randomly selected weeks test the hypothesis that there is no difference.

Statistic:	n	$\bar{x}$	$\sum(x-\bar{x})^2$
Finance	40	2503	2180.4
Legal	40	2510	2291.6

Non-normal distributions or unequal variances

What if some of the conditions required by the *t*-test are not met, e.g. the variances of the samples are different, or one or both of the samples do not appear to be normally distributed? The most important rule to learn is: **never use an inappropriate test!** If you cannot justify applying a *t*-test, look for an alternative, or consult an experienced statistician.

If the variances for two normally distributed samples are unequal but *known*, i.e.

$$X \sim N(\mu_x, \sigma_x^2) \quad \text{and} \quad Y \sim N(\mu_y, \sigma_y^2)$$

then

$$\bar{X} - \bar{Y} \sim N\left(\mu_x - \mu_y, \frac{\sigma_x^2}{m} + \frac{\sigma_y^2}{n}\right)$$

and so a suitable test statistic which will have a standard normal distribution under $H_0: \mu_x = \mu_y$ is

$$Z = \frac{\bar{X} - \bar{Y}}{\sqrt{\sigma_x^2/m + \sigma_y^2/n}}$$

If the variances are unknown, the problem is more difficult. It is known as the **Behrens-Fisher problem** and is discussed at length in more advanced text books.

For large samples (*m* and *n* both at least 30, say) we can apply the above procedure as an approximation, substituting s_x and s_y for σ_x and σ_y.

5.8 Non-parametric hypothesis tests

If we have any doubts about assuming that the samples are normally distributed, the mathematical justification for many hypothesis tests, including the *t*-test, breaks down, and we must turn to alternative, but less powerful, tests. These tests, which do not rely on a particular underlying distribution, are called **distribution-free** or **non-parametric** tests, and you will find some very large text books devoted entirely to these procedures. Here we shall look at two only; a third has already been discussed in Section 5.3.

The Wilcoxon signed-ranks test

This is a non-parametric alternative to the matched-pairs *t*-test of Section 5.6. As before, we first calculate the **differences** $d_i = Y_i - X_i$. This test tests the hypothesis that the median of the distribution of the d_i is zero – if the distribution of d_i is symmetrical, then the results of the test can also be interpreted in terms of $E(d_i) = E(Y_i) - E(X_i)$.

To perform the test, we rank in order the d_i from smallest to largest, without regard to sign. Then we sum the ranks corresponding to **positive** d_i to produce the Wilcoxon statistic T^+. Under the null hypothesis that the population distributions from which the matched pairs were drawn are identical, T^+ will, for large samples, be approximately normally distributed. We can calculate the mean and variance of T^+ from the ranks r_i assigned to *all* of the d_i:

$$E(T^+) = \frac{1}{2}\sum_{i=1}^{n} r_i \quad \text{and} \quad Var(T^+) = \frac{1}{4}\sum_{i=1}^{n} r_i^2$$

If there are no tied ranks (observations where the d_i have the same absolute value) the mean and variance are $\frac{1}{4}n(n+1)$ and $\frac{1}{24}n(n+1)(2n+1)$, respectively, as generally quoted in text books. However, if there are ties, we assign the *average* of the relevant ranks to all values in the tie, e.g. differences -0.2 0.5 -0.5 7 would receive ranks 1 2.5 2.5 4; and 4.2 -4.2 4.2 5 would receive ranks 2 2 2 4. This procedure has no effect on $E(T^+)$, but $Var(T^+)$ is changed. Finally note that T^+ can take only values that are multiples of 0.5, so a continuity correction is appropriate.

Example 7

Apply Wilcoxon's signed-ranks test to the data of Example 4.

The differences, and corresponding ranks, are:

Differences	10	2	23	8	−2	13	15	10	7	12
Ranks	5.5	1.5	10	4	1.5	8	9	5.5	3	7

T^+ is calculated by summing the ranks assigned to *positive* differences, so $T^+ = 53.5$.

$$E(T^+) = \frac{1}{2}\sum_{i=1}^{n} r_i = \frac{55}{2} = 27.5 \quad \text{and} \quad Var(T^+) = \frac{1}{4}\sum_{i=1}^{n} r_i^2 = \frac{384}{4} = 96$$

So

$$P(T^+ \geq 53.5) = 1 - \Phi\left(\frac{53.25 - 27.5)}{\sqrt{96}}\right) = 1 - \Phi(2.6281) = 0.00429$$

Note the continuity correction – $P(T^+ = 53.5)$ is represented by the area under the normal curve between 53.25 and 53.75.

Without assuming that the differences have a normal distribution, we can reject H_0 with significance probability 0.00429. Compare this with the significance probability of 0.00075 achieved by the *t*-test, demonstrating the extra **power** available if the normality assumption is justified.

The calculations may be performed in Minitab. With the original data in c1 and c2, calculate the differences (c3) and use *wtest*:

```
MTB > let c3=c2-c1
MTB > wtest c3;
SUBC> alternative 1.

TEST OF MEDIAN = 0.000000 VERSUS MEDIAN G.T. 0.000000

            N FOR   WILCOXON              ESTIMATED
        N   TEST    STATISTIC   P-VALUE   MEDIAN
C3      10   10      53.5        0.005    10.00
```

Note the use of the ALTERNATIVE subcommand to indicate a one-tailed test.

EXERCISES ON 5.8

1. Perform a Wilcoxon signed-ranks test on the chlorination data of Exercise 2 in section 5.6. How does the significance level compare with the one calculated for the paired t-test?
2. Using the cholesterol data (Dataset 1, Appendix A) take a random sample of 20 cases (in other words, select 20 numbers at random in the range 1–118, and look up the data corresponding to those values – you may decide to use a computer to do this, or a table of random numbers, or just put the 118 numbers in a hat and draw 20 out!). When you have selected your 20 cases, perform a Wilcoxon signed-ranks test on them to determine whether there is a significant change in (a) their cholesterol levels (*cholb* – *chola*), and (b) their weights (*wtb* – *wta*). Compare your results with those in Exercise 3 in Section 5.6, and discuss the effect of (i) changing from a t-test to Wilcoxon, (ii) reducing the size of the sample from 118 to 20.

The Mann–Whitney test

Also known as the **Wilcoxon rank sum test**, this is a non-parametric alternative to the two-sample t-test of Section 5.7. It does not require the data to be either normally or symmetrically distributed. The approach presented here is slightly non-standard, but somewhat less mysterious than that adopted by many texts.

Given the two samples, $X_1, X_2, \ldots, X_m$ and $Y_1, Y_2, \ldots, Y_n$, we pool the two datasets into one and then rank them. Then we calculate S, the sum of the ranks of the X data.

For example, X values: 1 2 3; Y values: 1 2 3 4.

Data Source	x	y	x	y	x	y	y
Ordered data	1	1	2	2	3	3	4
Ranks	1.5	1.5	3.5	3.5	5.5	5.5	7

S = sum of ranks of x data = $1.5 + 3.5 + 5.5 = 10.5$.

Now we recognise the fact that if the X and Y data have different means, then the sum of the ranks of the X data should be greater or less than what we would expect if the ranks used in the sum were a random sample **without replacement** from all the ranks.

128 Statistics

For small samples the actual distribution of S can be worked out under the assumption that the ranks in the sum have been chosen at random without replacement. For moderate to large samples, the distribution of S is approximately normal, with mean and variance calculated on the same assumptions, as follows.

Let $\bar{r}$ and s_r^2 denote the sample mean and sample variance of all the ranks. (For our example above, $\bar{r} = 4$ and $s_r = 2.102$.) Then

$$E(S) = m\bar{r} \quad \text{and} \quad Var(S) = ms_r^2\left(1 - \frac{m}{m+n}\right)$$

The first is obvious – remember that we are dealing with a sum of m ranks chosen at random from a sample of $m + n$ ranks with mean $\bar{r}$. The second is not so obvious – the factor $1 - m/(m + n)$ needs to be there because we are sampling the ranks without replacement (see Appendix B for the mathematical justification of this factor), otherwise the variance would have been the sum of the variances (ms_r^2).

Example 8

Forty patients at a hospital require operations to improve their eyesight. Two consultants at the hospital advocate different surgical procedures for the operations, both claiming that their method leads to a greater improvement in eyesight. They devise an experiment in which each operates on 20 of the patients, the allocation being at random, and an independent specialist measures the improvement in eyesight of each patient. The results are as follows (a higher score in the test indicates a greater improvement in eyesight). Use a Mann–Whitney test to determine whether either consultant's claim is correct.

Consultant A	2.8	2.3	2.5	2.9	3.4	2.3	3.5	3.8	3.5	2.5
	3.3	3.6	3.3	4.0	2.7	2.3	2.9	3.3	2.8	2.6
Consultant B	3.2	3.6	3.2	3.1	4.2	4.4	3.4	3.3	3.3	3.5
	3.5	3.4	3.5	2.9	4.3	3.9	3.0	4.4	3.3	3.4

First the combined dataset is ordered, and appropriate ranks are assigned:

Consultant	A	A	A	A	A	A	A	A	A	A
Improvement	2.3	2.3	2.3	2.5	2.5	2.6	2.7	2.8	2.8	2.9
Rank	2	2	2	4.5	4.5	6	7	8.5	8.5	11

Consultant	A	B	B	B	B	B	A	A	A	B
Improvement	2.9	2.9	3.0	3.1	3.2	3.2	3.3	3.3	3.3	3.3
Rank	11	11	13	14	15.5	15.5	19.5	19.5	19.5	19.5

Consultant	B	B	A	B	B	B	A	A	B	B
Improvement	3.3	3.3	3.4	3.4	3.4	3.4	3.5	3.5	3.5	3.5
Rank	19.5	19.5	24.5	24.5	24.5	24.5	29	29	29	29

Consultant	B	A	B	A	B	A	B	B	B	B
Improvement	3.5	3.6	3.6	3.8	3.9	4.0	4.2	4.3	4.4	4.4
Rank	29	32.5	32.5	34	35	36	37	38	39.5	39.5

The sum of the ranks associated with consultant A is 310.5.

$$\sum_{i=1}^{40} r_i = 820 \quad \text{so} \quad \bar{r} = \frac{820}{40} = 20.5$$

Therefore

$$E(S) = m\bar{r} = 20 \times 20.5 = 410.$$

$$\sum_{i=1}^{40} r_i^2 = 22101 \quad \text{so} \quad s_r^2 = \frac{22101 - 820^2/40}{39} = 135.667$$

Therefore

$$Var(S) = ms_r^2\left(1 - \frac{m}{m+n}\right) = 20 \times 135.667 \times 0.5 = 1356.67$$

$$S \sim N(410, 1356.67)$$

so

$$P(S \leq 310.5) = \Phi\left(\frac{310.75 - 410}{\sqrt{1356.67}}\right) = \Phi(-2.69) = 0.0036$$

This is the probability of the one tail. There are equally extreme values in the other tail, also with probability 0.0036, so the significance probability is $2 \times 0.0036 = 0.0072$.

Using Minitab:

```
MTB > mann-whitney c1 c2

Mann-Whitney Confidence Interval and Test

C1          N =  20      Median =    2.9000
C2          N =  20      Median =    3.4000
Point estimate for ETA1-ETA2 is    -0.6000
95.0 pct c.i. for ETA1-ETA2 is (-0.9000,-0.1000)
W = 310.5
Test of ETA1 = ETA2  vs.   ETA1 n.e. ETA2 is significant at 0.0074
The test is significant at 0.0072 (adjusted for ties)
```

Most text books define the **Mann–Whitney U-statistic**:

$$U = \text{sum of ranks of } X \text{ data} - \frac{1}{2}m(m+1)$$

The problem with this is that the quoted mean and variance do not cater for the situation when there are tied ranks – the approach presented above is completely general and requires only that there is sufficient data to justify the assumption of normality for the distribution of the sum of the X ranks.

EXERCISE ON 5.8

3. The managers of a large firm are investigating patterns of absences of their workforce. They compare total absences for 30 working days during the winter months of the current year with a similar set of figures taken five years ago.

Their data is listed below. Use a Mann–Whitney test to determine whether absenteeism appears to be getting worse. If you were in charge of this study, what other information would you want to collect?

Current year	38 28 34 42 35 32 28 31 35 33 39 35 34 36 38 36 33 40 40 33 36 38 40 35 33 36 41 36 41 35
Five years ago	27 36 33 35 35 32 34 32 30 36 34 30 31 32 30 30 33 35 31 35 32 31 35 34 34 35 31 34 39 35

5.9 Tests for proportions

Sometimes we have a theory which is best expressed by predicting the **proportion** or **percentage** of individuals in a population which have a particular attribute. For example, in a botanical experiment we may predict that 30% of seeds will germinate. Call the proportion of seeds germinating p. Then our null hypothesis would be H_0: $p = 0.3$. Suppose in an actual experiment with 100 seeds we observe that 25 germinate. We now have an *estimate* of the proportion $\hat{p} = 0.25$.

Of course, $\hat{p}$ is a random variable – if we repeat the experiment we will probably get a different value of $\hat{p}$. So is our null hypothesis value of 0.3 acceptable? We need to know the *distribution* of $\hat{p}$.

If we assume that the individuals (seeds) are independent and each has the same probability of having the characteristic (germinating), the *number* of 'successes' (germinating seeds) will have a **binomial** distribution with parameters $n = 100$ (the sample size) and p (the true proportion of seeds germinating). Call this number X, i.e. $X \sim Bi(n, p)$.

Provided n is large, we can use a normal approximation to this binomial distribution, with mean np and variance $np(1 - p)$. So approximately $X \sim N(np, np(1 - p))$.

Our proportion $\hat{p}$ is simply X/n, which is a linear function. So (still approximately) $\hat{p}$ will be normally distributed with mean $\frac{1}{n}E(X)$ and variance $(1/n)^2 Var(X)$, i.e.

$$\hat{p} \sim N\left(p, \frac{p(1-p)}{n}\right) \quad \text{or} \quad \frac{\hat{p} - p}{\sqrt{p(1-p)/n}} \sim N(0, 1)$$

We now have an approximation for the distribution of $\hat{p}$, based on the true value of p. If H_0 is true, then $p = 0.3$ and $\hat{p} \sim N(0.3, 0.0021)$. What is the probability of obtaining a $\hat{p}$ value as extreme as 0.25?

$$P(\hat{p} < 0.25) = \Phi\left(\frac{0.25 - 0.3}{\sqrt{0.0021}}\right) = \Phi(-1.0911) = 0.1376$$

There are equally extreme values in the region $\hat{p} > 0.35$, which also has probability 0.1376, so values as extreme as 0.25 have total probability $2 \times 0.1376 = 0.2752$. In other words, if the true value of p is 0.3, we have better than 1 chance in 4 of obtaining such a value, so we cannot reject H_0: $p = 0.3$.

The difference between two proportions

Now suppose that we observe two proportions $\hat{p}_1$ and $\hat{p}_2$ in separate experiments, with different experimental procedures or conditions. We will often want to test whether the population proportions p_1 and p_2 are significantly different.

From our calculations above,

$$\hat{p}_1 \sim N\left(p_1, \frac{p_1(1-p_1)}{n_1}\right)$$

and

$$\hat{p}_2 \sim N\left(p_2, \frac{p_2(1-p_2)}{n_2}\right)$$

so, differencing these two normally distributed random variables,

$$\hat{p}_1 - \hat{p}_2 \sim N\left(p_1 - p_2, \frac{p_1(1-p_1)}{n_1} + \frac{p_2(1-p_2)}{n_2}\right)$$

Under $H_0: p_1 = p_2$,

$$\hat{p}_1 - \hat{p}_2 \sim N\left(0, p(1-p)\left(\frac{1}{n_1} + \frac{1}{n_2}\right)\right)$$

where p is the true (common) proportion.

We must now *estimate* p, using

$$\hat{p} = \frac{n_1 \hat{p}_1 + n_2 \hat{p}_2}{n_1 + n_2}$$

(the *pooled* proportion).

Example 9

In an experiment to test the effectiveness of two insecticides, 80 insects are sprayed with insecticide 1, and 45 die, whilst 70 insects are sprayed with insecticide 2 and 47 die. Is there evidence that one insecticide is significantly better than the other?

The estimated proportions are $\hat{p}_1 = 45/80 = 0.5625$ and $\hat{p}_2 = 47/70 = 0.6714$. These are random variables which are estimates of the true (population) proportions p_1 and p_2 (the 'long-term' average proportion that would be killed by insecticides 1 and 2 respectively). To decide whether p_1 and p_2 are really different, we put forward the null hypothesis that they are the same, $H_0: p_1 = p_2$. If they are the same, our best estimate of this common value is

$$\hat{p} = \frac{n_1 \hat{p}_1 + n_2 \hat{p}_2}{n_1 + n_2} = \frac{80 \times 0.5625 + 70 \times 0.6714}{80 + 70} = 0.613$$

(We can also obtain $\hat{p}$ from $(45 + 47)/(80 + 70)$.) So

$$\hat{p}_1 - \hat{p}_2 \sim N\left(0, \hat{p}(1-\hat{p})\left(\frac{1}{n_1} + \frac{1}{n_2}\right)\right) \quad \text{or} \quad \hat{p}_1 - \hat{p}_2 \sim N(0, 0.00635)$$

Our observed value of $\hat{p}_1 - \hat{p}_2$ is $0.5625 - 0.6714 = -0.1089$ and

$$P(\hat{p}_1 - \hat{p}_2 < -0.1089) = \Phi\left(\frac{-0.1089}{\sqrt{0.00635}}\right) = \Phi(-1.3666) = 0.086$$

For a two-tailed test, we double this value to obtain a significance probability of 0.172. This observed difference of proportions is therefore quite an acceptable value under the hypothesis that there is no difference between the two insecticides.

EXERCISE ON 5.9

1. In a study on prosperity in different areas of a city, 100 houses were selected at random in each of two districts. In district A, thought by the researchers to be the more prosperous of the two, 60 houses had double-glazing and 35 had burglar alarms. In district B, 47 had double-glazing and 22 had burglar alarms. Perform suitable hypothesis tests to determine whether there is a significant difference between the areas on the basis of these two measures.

5.10 The distribution of the sample variance

The chi-squared distribution

If we take n independent random variables, each with an $N(0, 1)$ distribution, square them and add them together, the resulting random variable will have a distribution called the **chi-squared distribution** with n degrees of freedom (usually written $\chi^2(n)$). Like the t distribution, this is really a **family** of distributions, with a different shape for each number of degrees of freedom. Unlike t, the chi-squared distribution is asymmetrical. In statistical tables, percentage points for the chi-squared distribution are tabulated for various number of degrees of freedom and for fixed significance levels, such as 5% and 1%. Because of the asymmetry, percentiles at either extreme are tabulated, e.g. the 1st and 99th percentiles. The graphs of the probability density functions for different values of n are shown in Fig 5.3.

If we draw a random sample of size n from a population of normally distributed values, whose true variance is σ^2, and then calculate the sample variance s^2, the statistic $(n-1)s^2/\sigma^2$ will have a χ^2 distribution with $n-1$ degrees of freedom.

Confidence intervals and hypothesis tests for the population variance

From the above, we can say that

$$P(\chi^2_{0.025} < (n-1)s^2/\sigma^2 < \chi^2_{0.975}) = 0.95$$

so we obtain a 95% confidence interval for σ^2 by manipulating these inequalities:

$$(n-1)s^2/\chi^2_{0.975}, (n-1)s^2/\chi^2_{0.025}$$

For example, suppose we calculate a sample variance of 5.7 from a sample of size 30. From statistical tables, $\chi^2_{0.025}$ with 29 degrees of freedom is 16.047 and $\chi^2_{0.975}$

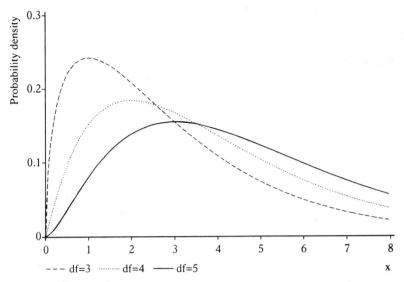

Fig 5.3 χ^2 probability density functions with $n = 1, 2$ and 3.

with 29 degrees of freedom is 45.722, so a 95% confidence interval for σ^2 is

$$\left(\frac{29 \times 5.7}{45.722}, \frac{29 \times 5.7}{16.047}\right) = (3.615, 10.301)$$

Example 10

A soft drinks firm's quality control department requires that the quantity of its product dispensed by an automatic vending machine should be distributed with a standard deviation of no more than 2 ml. A hundred units are tested and the calculated standard deviation is 2.5 ml. Is this sufficient evidence to reject the machine?

We need to find out how likely it is that we would observe a sample value of 2.5 when the population value is no more than 2. First we shall convert to measures of the *variance*: our null hypothesis should be H_0: $\sigma^2 = 4$, and we have observed $s^2 = 6.25$:

$$\frac{(n-1)s^2}{\sigma^2} = \frac{99 \times 6.25}{4} = 154.6875$$

Under H_0 this should have a χ^2 distribution with 99 degrees of freedom. Using Minitab:

```
MTB > cdf 154.6875;
SUBC> chisquare 99.
    154.6875    0.9997
```

So the probability of observing a value as big as 154.6875 is $1 - 0.9997 = 0.0003$. This probability should be doubled to allow for equally extreme small values of s^2, making a significance probability of 0.0006, i.e. it is extremely unlikely that σ^2 is really 4.

The F distribution

If Y has a χ^2 distribution with m degrees of freedom and Z has a χ^2 distribution with n degrees of freedom, and Y and Z are independent, then the quotient $(Y/m)/(Z/n)$ has a distribution called the **F distribution** with m and n degrees of freedom (written $F(m, n)$). Recalling that χ^2 is the distribution of the sum of squares of normal random variables, we see that F is the distribution of the ratio of two sums of squares.

In statistical tables, critical points of the F distribution are tabulated, usually with one significance level per page because both degrees of freedom must be listed. Note that the order of the degrees of freedom matters – $F(3, 5)$ is a different distribution from $F(5, 3)$.

The variance-ratio test

Suppose we draw a random sample $X_1, X_2, \ldots, X_m$ from a normal distribution with variance σ_x^2 and a random sample $Y_1, Y_2, \ldots, Y_n$ from a normal distribution with variance σ_y^2, and calculate the sample variances s_x^2 and s_y^2. Recall that if s^2 is the sample variance of a random sample of size n from a normal distribution with variance σ^2, $(n-1)s^2/\sigma^2$ has a χ^2 distribution with $n-1$ degrees of freedom. It follows that $(s_x^2/\sigma_x^2)/(s_y^2/\sigma_y^2)$ has an F distribution with $m-1$ and $n-1$ degrees of freedom.

If $\sigma_x^2 = \sigma_y^2$ we can deduce that

$$\frac{s_x^2}{s_y^2} \sim F(m-1, n-1)$$

So under the null hypothesis H_0: $\sigma_x^2 = \sigma_y^2$ (i.e. under the assumption that the population variances for X and Y are equal) the statistic s_x^2/s_y^2 will have an F distribution with $m-1$ and $n-1$ degrees of freedom. The **variance-ratio test** is performed by calculating the ratio s_x^2/s_y^2 and determining the corresponding significance probability for the F distribution. It is conventional to make the larger of s_x^2 and s_y^2 the numerator. Care must then be exercised in calculating the significance level to reflect whether the test is two-tailed or one-tailed.

Example 11

Ten 1 m lengths of yarn (controls) are stretched until they break, and the extension in cm is recorded. A batch of yarn is washed, and a sample of nine 1 m lengths is similarly tested. The results are recorded below. Perform a variance-ratio test to determine whether there is a significant difference between the variances.

Washed (X)	1.3	1.6	0.8	1.2	1.4	1.1	1.3	1.4	0.9	
Control (Y)	1.5	1.5	1.0	2.0	1.7	1.3	0.4	1.2	1.2	1.3

For the washed yarn, $m = 9$, $s_x^2 = 0.06444$. For the controls, $n = 10$, $s_y^2 = 0.18322$. So, taking the larger as the numerator, $s_y^2/s_x^2 = 2.84$. The variance of the control sample is almost three times the variance of the treated sample. Is this significant?

Using Minitab:

```
MTB > cdf 2.8431;
SUBC> F 9 8.
      2.8431    0.9220
```

So the probability of such a large ratio is $1 - 0.9220 = 0.078$. Generally, variance-ratio tests are two-tailed, since we do not usually have any reason to believe that one particular sample will have a larger variance. So this probability must be doubled to 0.156. In other words, we would expect such an extreme result in 15.6% of such tests, so even a ratio of nearly 3 is not really very significant.

When we consider performing a two-sample t-test (see Section 5.7) to check whether the means of two populations are different, we know that one of the requirements is that the variances should be equal. Because of this, many researchers use the variance-ratio test to check the validity of the t-test. For fairly large samples this is a reasonable practice, but for small samples there has to be a large ratio between the variances before the variance-ratio test becomes significant even at the 5% significance level. For example, when both samples are of size 10, the ratio between the sample variances must exceed 4.03 for 5% significance.

EXERCISE ON 5.10

1. In a study of the bacterial content of the water off a bathing beach, a random selection of sampling days gave the following results for the quantity of bacteria present. The measurement is the log coliform density per ml and the days have been classified as either 'sunny' or 'cloudy'. The researchers believe that there will be more variation in the measurements on sunny days. Perform a variance-ratio test on the data to determine whether they are right.

Sunny	4.4	3.9	3.4	3.2	3.9	4.1	3.8	3.7	
Cloudy	3.0	3.7	3.0	3.4	2.0	3.1	3.4	3.0	3.4

   A statistician in their office tells them 'you need about 50 observations of each type of day before you can prove anything with that sort of data'. Do you think the statistician is right? What sort of assumptions is the statistician making?

5.11 Hypothesis tests in regression and correlation

In Section 2.10 we calculated the sums of squares S_{xx}, S_{yy} and S_{xy} in order to estimate the correlation coefficient, and later to estimate the coefficients in the simple linear regression $Y = \beta_0 + \beta_1 X$. These sums of squares are of interest in their own right.

S_{yy} summarises **the amount of variation in the y data** without regard to its relationship with the x data – if we want to fit a horizontal line ($\beta_1 = 0$) through the data, then the least-squares line is the horizontal line at level $\bar{y}$, and the least-squares value is S_{yy}.

$S_{yy} - \hat{\beta}_1 S_{xy}$ is the residual sum of squares, summarising the amount of variation in the (x, y) data **about the fitted line**. The difference $\hat{\beta}_1 S_{xy}$ summarises the amount of variation in the y data **explained by the regression of y on x**. We can record this in a table (Table 5.2):

Table 5.2 Analysis of variance.

Due to regression of y on x	$\hat{\beta}_1 S_{xy}$	df = 1
Residual sum of squares	$S_{yy} - \hat{\beta}_1 S_{xy}$	df = $n - 2$
Total	S_{yy}	df = $n - 1$

Testing for a linear relationship

We obtain the total sum of squares S_{yy} by estimating one parameter, so S_{yy} has $n - 1$ degrees of freedom. The residual sum of squares involves estimating two parameters and has $n - 2$ degrees of freedom. The difference between the two, the sum of squares due to the regression, $\hat{\beta}_1 S_{xy}$, therefore has one degree of freedom.

To test for a linear relationship, we divide the sums of squares by their degrees of freedom and produce a quotient:

$$\frac{\hat{\beta}_1 S_{xy}/1}{(S_{yy} - \hat{\beta}_1 S_{xy})/(n - 2)}$$

If $\beta_1 = 0$, this will have an F distribution with 1 and $n - 2$ degrees of freedom.

Example 12

For the cholesterol data (Dataset 1, Appendix A), perform a simple linear regression of *cholb* on *age* and test for a linear relationship.

First we shall use Minitab to plot the data:

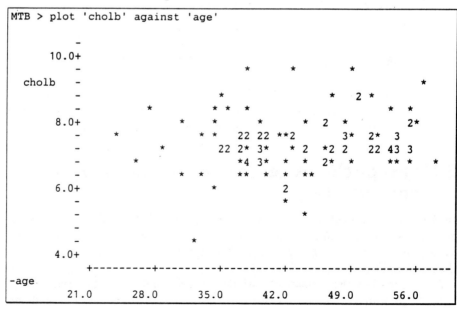

It is not at all obvious that this data follows any sort of linear relationship.

Calculate the sums of squares: $S_{xx} = 7062.20$, $S_{yy} = 78.20$, $S_{xy} = 116.93$. So $\hat{\beta}_1 = 116.93/7062.20 = 0.01656$ and $\hat{\beta}_0 = \bar{y} - \hat{\beta}_1 \bar{x} = 7.3525 - 0.01656 \times 43.712 = 6.629$, i.e.

$$cholb = 6.629 + 0.01656 \times age$$

Table 5.3 Analysis of variance.

Due to regression	$\hat{\beta}_1 S_{xy} = 0.01656 \times 116.93 = 1.936$
Residual sum of squares	$S_{yy} - \hat{\beta}_1 S_{xy} = 78.20 - 1.936 = 76.264$
Total	$S_{yy} = 78.20$

The F-statistic is, using Table 5.2,

$$\frac{\hat{\beta}_1 S_{xy}/1}{(S_{yy} - \hat{\beta}_1 S_{xy})/(n-2)} = \frac{1.936}{76.264/116} = 2.94$$

The significance probability of this value is 0.089.

All of these results are calculated by Minitab's REGRESS command, as in Fig 5.4.

```
MTB > regress 'cholb' on 1 predictor 'age'

The regression equation is
cholb = 6.63 + 0.0166 age

Predictor       Coef        Stdev       t-ratio      p
Constant        6.6287      0.4283      15.48        0.000
age             0.016558    0.009649    1.72         0.089

s = 0.8108      R-sq = 2.5%     R-sq(adj) = 1.6%

Analysis of Variance

SOURCE          DF          SS          MS           F         p
Regression      1           1.9361      1.9361       2.94      0.089
Error           116         76.2648     0.6575
Total           117         78.2010
```

Fig 5.4 Regression of cholb on age.

Although not significant at 5%, this significance probability is small enough to suggest that there is some linear relationship between the variables. The value of $\hat{\beta}_1$ suggests that for an increase in *age* of 1 year *cholb* may be expected to increase by 0.017.

Estimating the variance

An important assumption for the method of least squares to be appropriate is that the variances of all the observations are the same. So for the *i*th point we can write our regression model as $Y_i = \beta_0 + \beta_1 x_i + e_i$ where $Var(Y_i) = Var(e_i) = \sigma^2$. An estimate of σ^2 is obtained from the denominator of the F-statistic above:

$$\hat{\sigma}^2 = \frac{\text{residual sum of squares}}{n - 2}$$

If the regression model holds, then $(n - 2)\hat{\sigma}^2/\sigma^2$ has a $\chi^2(n - 2)$ distribution.

Confidence intervals and hypothesis tests

Because $\hat{\beta}_0$ and $\hat{\beta}_1$ are dependent on the Y_i (the y data regarded as random variables) they have normal distributions, with means equal to β_0 and β_1, respectively. We can calculate their variances.

To find the variance of $\hat{\beta}_1$:

$$\hat{\beta}_1 = \frac{S_{xy}}{S_{xx}} = \frac{\sum Y_i(x_i - \bar{x})}{S_{xx}} = \sum b_i Y_i \quad \text{where} \quad b_i = \frac{(x_i - \bar{x})^2}{S_{xx}}$$

Therefore

$$Var(\hat{\beta}_1) = \sum b_i^2 Var(Y_i) = \sigma^2 \sum b_i^2 = \sigma^2 \sum \frac{(x_i - \bar{x})^2}{S_{xx}^2} = \frac{\sigma^2}{S_{xx}}$$

To find the variance of $\hat{\beta}_0$:

$$\hat{\beta}_0 = \bar{Y} - \hat{\beta}_1 \bar{x}$$

It is not immediately obvious, but $\bar{Y}$ and $\hat{\beta}_1$ are statistically independent. So

$$Var(\hat{\beta}_0) = Var(\bar{Y}) + \bar{x}^2 Var_1(\hat{\beta}_1) = \frac{\sigma^2}{n} + \frac{\bar{x}^2 \sigma^2}{S_{xx}} = \sigma^2 \left(\frac{1}{n} + \frac{\bar{x}^2}{S_{xx}} \right)$$

Hence hypotheses about β_0 or β_1 can be tested by calculating appropriate t-statistics:

$$t = \frac{\hat{\beta}_0 - \beta_0}{\hat{\sigma}\sqrt{1/n + \bar{x}^2/S_{xx}}} \quad \text{and} \quad t = \frac{\hat{\beta}_1 - \beta_1}{\hat{\sigma}/\sqrt{S_{xx}}}$$

To find the variance of $\hat{\beta}_0 + \hat{\beta}_1 x$ (which estimates the value of the true regression line at point x):

$$Var(\hat{\beta}_0 + \hat{\beta}_1 x) = Var[\bar{Y} + \hat{\beta}_1(x - \bar{x})] = \frac{\sigma^2}{n} + \frac{\sigma^2(x - \bar{x})^2}{S_{xx}} = \sigma^2 \left(\frac{1}{n} + \frac{(x - \bar{x})^2}{S_{xx}} \right)$$

However, if we want to predict what a fresh Y observation at x will be, then our best estimate will be the same quantity, $\hat{\beta}_0 + \hat{\beta}_1 x$, but the variance associated with this prediction will be

$$\sigma^2 \left(1 + \frac{1}{n} + \frac{(x - \bar{x})^2}{S_{xx}} \right)$$

• Example 13

Using the cholesterol data (Dataset 1, Appendix A), regress *wta* on *wtb*. Calculate 95% confidence intervals for (a) the variance σ^2, (b) the 'slope' coefficient β_1, (c) the mean value of *wta* for *wtb* = 100, (d) a prediction interval for a fresh observation at $wtb = 100$. Test the hypothesis that $\beta_1 = 1$ at the 5% significance level.

The sums of squares are: $S_{xx} = 23624.01$, $S_{yy} = 22881.97$, $S_{xy} = 22599.88$. So $\hat{\beta}_1 = 0.9566$, $\hat{\beta}_0 = 1.379$, RSS $= 1261.8$.

1. $\hat{\sigma}^2 = \dfrac{1261.8}{116} = 10.88 \quad \chi^2_{0.025}(\text{df} = 116) = 88.08 \quad \chi^2_{0.975}(\text{df} = 116) = 147.7$

Therefore, with probability 0.95, $88.08 \leq 116\hat{\sigma}^2/\sigma^2 \leq 147.7$, or $116\hat{\sigma}^2/147.7 \leq \sigma^2 \leq 116\hat{\sigma}^2/88.08$. Substituting for $\hat{\sigma}^2$, a 95% confidence interval for σ^2 is $(8.54, 14.33)$.

2. $\hat{\beta}_1 = 0.9566 \quad \dfrac{\hat{\sigma}^2}{S_{xx}} = \dfrac{10.88}{23624.01} = 0.00046055 \quad t_{0.975}(116) = 1.98$

Therefore a 95% confidence interval for $\hat{\beta}_1$ is

$$0.9566 \pm 1.98\sqrt{0.00046055} = (0.914, 0.999)$$

To test the hypothesis that $\beta_1 = 1$, construct the t-statistic:

$$t = \dfrac{\hat{\beta}_1 - 1}{\sqrt{\hat{\sigma}^2/S_{xx}}} = \dfrac{0.9566 - 1}{\sqrt{10.88/23624.01}} = -2.0223$$

From the computer, the one-tailed significance probability is $1 - 0.9773 = 0.0227$. For a two-tailed test this is, doubled to 0.045. So the hypothesis that $\beta_1 = 1$ is *rejected* at the 5% significance level.

3. A confidence interval for the mean value of wta for $wtb = 100$:

Predicted mean $wta = \hat{\beta}_0 + 100 \times \hat{\beta}_1 = 97.039$.
   The standard error of this prediction is

$$\hat{\sigma}\sqrt{\dfrac{1}{118} + \dfrac{(100 - 75.93)^2}{23624.01}} = 0.599$$

Therefore a 95% confidence interval for

$$E(wta|wtb = 100) = 97.039 \pm 0.599 \times 1.98 = (95.85, 98.23)$$

4. A prediction interval for a fresh observation at $wtb = 100$:

Predicted value of $wta = \hat{\beta}_0 + 100 \times \hat{\beta}_1 = 97.039$.

Estimated standard error of a new observation at $wtb = 100$ is

$$\hat{\sigma}\sqrt{1 + \dfrac{1}{118} + \dfrac{(100 - 75.93)^2}{23624.01}} = 3.352$$

Therefore a 95% confidence interval is $97.039 \pm 3.352 \times 1.98 = (90.4, 103.7)$.

EXERCISE ON 5.11

1. The following data summarises, for 15 selected geographical areas throughout the United States, the average daily maximum temperature and the average

percentage of sunshine for the month of May, based on a 30 year period from 1951 to 1980.

Temperature	85 92 69 66 71 71 85 70 73 85 63 74 75 72 87
Sunshine	65 93 66 72 64 58 70 60 61 62 55 56 64 61 89

X = temperature, Y = sunshine.

$$\sum x = 1138 \quad \sum y = 996 \quad \sum x^2 = 87390 \quad \sum y^2 = 67838$$

$$\sum xy = 76466$$

(a) Calculate the least-squares estimates of the regression line of percentage sunshine on temperature.
(b) Test the hypothesis that the slope of the regression line is zero at the 5% level by producing both a t-statistic and an F-statistic. (Check that $t^2 = F$ and that the conclusions are identical.)
(c) Comment on whether you would be prepared to use this relationship for predicting sunshine for another month.
(d) Produce a scatter diagram of the data with the estimated regression line superimposed. In view of this plot, comment on whether you think the model (and your answer to (b)) is really appropriate.

Testing the significance of the linear correlation coefficient

When we want to decide whether two variables are linearly correlated, we can test the hypothesis that the regression coefficient β_1 is 0. Alternatively we can perform a test for the **population linear correlation coefficient**, ρ, which measures the linear relationship between two populations. The test statistic we use is based on the Pearson correlation coefficient, r, defined in Section 2.10. Note that

$$r = \frac{S_{xy}}{\sqrt{S_{xx}S_{yy}}} = \hat{\beta}_1 \sqrt{\frac{S_{xx}}{S_{yy}}}$$

The test of the hypothesis $\rho = 0$ is equivalent to the test that the regression line slope $\beta_1 = 0$. This has test statistic

$$t = \frac{\hat{\beta}_1}{\sqrt{\hat{\sigma}^2/S_{xx}}}$$

In terms of r,

$$\hat{\beta}_1 = r\sqrt{\frac{S_{yy}}{S_{xx}}}$$

and

$$\frac{\hat{\sigma}^2}{S_{xx}} = \frac{S_{yy} - \hat{\beta}_1 S_{xy}}{(n-2)S_{xx}} = \frac{S_{yy}}{(n-2)S_{xx}}\left(1 - \frac{S_{xy}^2}{S_{xx}S_{yy}}\right) = \frac{S_{yy}}{(n-2)S_{xx}}(1 - r^2)$$

So the test statistic

$$t = \frac{\hat{\beta}_1}{\sqrt{\hat{\sigma}^2/S_{xx}}} = \frac{r\sqrt{S_{yy}/S_{xx}}}{\sqrt{S_{yy}/S_{xx} \times (1-r^2)/(n-2)}} = r\sqrt{\frac{n-2}{1-r^2}}$$

has a t distribution with $n - 2$ degrees of freedom.

Example 14

IQ scores are scaled to have an average of 100 and standard deviation 15 for both men and women. The correlation between IQs of husbands and wives is about 0.5. In a large study of families, it was found that for the men whose IQ was 140 the average IQ of their wives was 120. Looking at the wives of IQ 120 in the study, is it fair to expect that the average IQ of their husbands will be greater than 120?

Since the standard deviations are equal, the regression line of wives' IQ given husbands' IQ is $100 + 0.5(x - 100)$. So when $x = 140$, the expected IQ of wives is 120. Similarly, the regression line of husbands' IQ given wives' IQ is $100 + 0.5(x - 100)$, so when $x = 120$, the expected husbands' IQ is 110.

Example 15

At birth 27 identical twins were separated and one (y data) raised by foster parents, and the other (x data) raised by natural parents. Their IQs were measured, with the following results:

	y	x
Mean	95.111	95.296
Standard deviation	16.082	15.735

Correlation coefficient $= 0.882$.

(a) Test the hypothesis $\rho = 0$.
(b) Compute the expected Y for a given $x = 70$.

(a) The t-statistic is $0.882 \times \sqrt{25/(1 - 0.882^2)} = 9.36$, which is clearly highly significant, so $\rho \neq 0$.
(b) Estimated $E(Y|x = 70) = 95.111 + 0.882 \times (16.082/15.735)(70 - 95.296) = 71.8$. Note that despite the fact that the mean of x is greater than the mean of y, this value is greater than 70 (though not by much because the correlation between y and x is so high).

EXERCISE ON 5.11

2. The following four datasets were published by Frank J. Anscombe in *American Statistician* (1973). Calculate the correlation coefficient for each set, and then produce scatter plots of each. This exercise is mandatory for anyone who really wants to understand correlation!

X_1	Y_1	X_2	Y_2	X_3	Y_3	X_4	Y_4
10.00	8.04	10.00	9.14	10.00	7.46	8.00	6.58
8.00	6.95	8.00	8.14	8.00	6.77	8.00	5.76
13.00	7.58	13.00	8.74	13.00	12.74	8.00	7.71
9.00	8.81	9.00	8.77	9.00	7.11	8.00	8.84
11.00	8.33	11.00	9.26	11.00	7.81	8.00	8.47
14.00	9.96	14.00	8.10	14.00	8.84	8.00	7.04
6.00	7.24	6.00	6.13	6.00	6.08	8.00	5.25
4.00	4.26	4.00	3.10	4.00	5.39	19.00	12.50
12.00	10.84	12.00	9.13	12.00	8.15	8.00	5.56
7.00	4.82	7.00	7.26	7.00	6.42	8.00	7.91
5.00	5.68	5.00	4.74	5.00	5.73	8.00	6.89

As a further exercise, calculate the estimates of the linear regression coefficients, and the residual sum of squares.

5.12 The chi-squared goodness-of-fit test

Much of statistics, and indeed science in general, has to do with devising appropriate, relatively simple models to represent a reality that is rarely simple! Whether we can believe our chosen model to be a true representation of the reality in question is debatable. Nevertheless we may wish to use the model for predictive purposes and to do so without checking that 'the model fitted the evidence/data' would be unacceptable. Hence the need for a 'goodness-of-fit' procedure to validate the chosen model. The χ^2 goodness-of-fit test may be applied in a wide variety of contexts, provided only that, on the basis of the model, the data can be classified into one of k categories with observed frequencies O_i and expected frequencies E_i calculated assuming the model to be true.

For example, the frequencies of each class interval of a histogram (the O_i) may be compared with expected frequencies (the E_i) calculated or deduced from a nominated distribution. In such a situation, an acceptable result from a goodness-of-fit test would mean that the data appeared to have the chosen distribution.

As we shall see later, another application of the χ^2 goodness-of-fit test would be to test whether two variables were statistically independent. The variables in question need not be ratio-scale variables and a typical application might, for example, test whether categorising people into different eye disorders was independent of their sex.

The statistic which measures how well the model fits the observations is

$$Q = \sum_{i=1}^{k} \frac{(O_i - E_i)^2}{E_i}$$

If the proposed model is correct, then this quantity has a distribution which approximates to a χ^2 distribution.

Hypothesis Testing 143

To determine the appropriate number of degrees of freedom in any specific application, we apply the rule

$$n = k - con - par$$

where

k = number of observed frequencies
con = number of constraints on the expected frequencies
par = number of parameters estimated from the data in order to calculate the expected frequencies.

Any unexpectedly large discrepancy between an O_i and its corresponding E_i will increase the value of Q, so the model should be rejected if Q is too large. The approximation to the χ^2 distribution is good if all the E_i are at least 5 and $k \geq 5$. If $k < 5$, larger E_i are preferable. If some of the E_i are less than 5, the categories should be grouped sensibly to ensure that all $E_i \geq 5$.

Example 16

Use the χ^2 goodness-of-fit test to determine whether the binomial distribution provides a good model for the computer failure data in Section 3.2.

The distribution fitted is binomial, with $n = 25$ (because there are 25 computers) and $p = 0.042$ (estimated from the data). The frequencies predicted by the model (which should not be rounded for this purpose) are given in Table 5.4.

Table 5.4 Probabilities and expected frequencies for Bi(25, 0.042).

Value	Probability	Expected frequency
0	0.342088	34.2088
1	0.374940	37.4940
2	0.197254	19.7254
3	0.066301	6.6301
4	0.015987	1.5987
5	0.002944	0.2944
6	0.000430	0.0430
7	0.000051	0.0051
8	0.000005	0.0005

The combined expected frequencies for the values 4–8 sum to less than 2, so we shall combine these with the frequency for 3. This leaves us only four categories, but all the expected frequencies are considerably more than 5, so the χ^2 approximation will be acceptable. The complete calculation is illustrated in Table 5.5. Note, however, that we have one overall constraint (the frequencies must sum to the total number of observations) and we have estimated one parameter (p for the binomial distribution) so we have $4 - 2 = 2$ degrees of freedom.

Table 5.5 The χ^2 calculation for the computer data.

Value	Probability	Expected frequency (e_i)	Observed frequency (o_i)	$\dfrac{(o_i - e_i)^2}{e_i}$
0	0.342088	34.2088	36	0.093789
1	0.374940	37.4940	34	0.325600
2	0.197254	19.7254	21	0.082361
3+	0.085718	8.5718	9	0.021391
				$Q = 0.523$

The 95th percentile of the χ^2 distribution with two degrees of freedom is 5.99, so clearly Q is not significant, and we can deduce that the observed frequencies are consistent with the proposed binomial model. (We have not, of course, *proved* that the model is correct, but our experiment has failed to uncover any evidence to the contrary.)

Example 17

A standard six-sided die with faces labelled with values 1–6 is thrown 60 times, and the following frequencies for the faces are obtained:

Face	1	2	3	4	5	6
Frequency	6	10	8	8	9	19

Is there evidence to support the hypothesis that the die is biased?

We shall use the χ^2 goodness-of-fit statistic to test the *null hypothesis* H_0: the die is fair. If H_0 is rejected, we can conclude that the die is biased.

Under H_0 we would expect the frequencies of all faces to be equal (in this case 10). So

$$Q = \frac{(6-10)^2}{10} + \frac{(10-10)^2}{10} + \frac{(8-10)^2}{10} + \frac{(8-10)^2}{10} + \frac{(9-10)^2}{10} + \frac{(19-10)^2}{10}$$
$$= 1.6 + 0 + 0.4 + 0.4 + 0.1 + 8.1 = 10.6$$

The significance probability for this value (χ^2 with five degrees of freedom) is 0.0599. We can deduce that there is some evidence to support the hypothesis that the die is biased, but it is not quite significant at the 5% level.

EXERCISE ON 5.12

1. A consignment of electrical switches is imported in 1000 boxes, each containing 10 switches. It is suspected that there is a manufacturing defect which would result in a serious malfunction of some of the switches, so a random sample of 500 boxes is selected and all 5000 switches are tested, with the following number of defective switches found per box:

Defectives	0	1	2	3	4	5	>5
Number of boxes	93	184	139	65	18	1	0

Let X be the number of defective switches in a box. Assume that the malfunction occurs independently for each switch with a constant probability p.
(a) What distribution would you expect random variable X to have?
(b) Use the above data to estimate the parameters of the distribution.
(c) Use your fitted distribution to calculate expected frequencies corresponding to those observed in the table.
(d) Use the χ^2 goodness-of-fit test to check whether the model seems to fit.

Testing for a fit that is too good

In most situations, we only consider rejecting the null hypothesis, that the proposed model is correct, if we observe values that differ from the expected values. This is why we use one-tailed tests. Suppose now that the die in Example 17 gave the following frequencies:

Face	1	2	3	4	5	6
Frequency	10	10	9	10	10	11

Our first reaction may be that this is a very good fit. We may then consider whether the fit is in fact *too good*. Is it possible for the fit to be too good? It is difficult to imagine how a real six-sided die could produce such results, except by chance, but what if the "experiment" was carried out on a computer? Suppose a sequence such as 1, 2, 3, 4, 5, 6 was produced rather too often. Our conclusion may then be that there is 'not enough randomness' – the fit is too good.

To test for this eventuality, we need the low percentiles of the χ^2 distribution. For a 5% one-tailed test to see if the fit is too good, we would use the 5th percentile, and for a two-tailed test (to test both extremes simultaneously) we would use percentiles 2.5 and 97.5.

For the data above, $Q = 0.2$. Percentile 2.5 of the $\chi^2(5)$ distribution is 0.831, and percentile 0.5 is 0.412, so the result is significant at the 1% significance level. There is considerable evidence that the fit is too good.

Some possible reasons for such an occurrence of 'too good a fit' are:

- many datasets were collected, and the 'best one' chosen;
- throws of the die occurring adjacent in time are highly correlated;
- the die model is just totally inappropriate – for example, if the throws of the die were deterministic as in the sequence 1 2 3 4 5 6 1 2 3 4 5 6 . . . .

Example 18

In an experiment on heredity a pure-breeding tall pea plant possessing coloured flowers is crossed with a dwarf plant possessing white flowers. Four different kinds of plant were produced: (1) tall with coloured flowers; (2) tall with white flowers; (3) dwarf with coloured flowers; (4) dwarf with white flowers. In all 160 plants were produced, and the observed numbers of different kinds of plant were 85, 32, 34 and 9, respectively. Mendel's theory predicts that the probability of obtaining a tall plant is 0.75 and, independently, the probability of obtaining coloured flowers is 0.75. Are the experimental results consistent with Mendel's theory?

Since the two attributes are independent,

$$P(\text{tall and coloured}) = \frac{3}{4} \times \frac{3}{4} = \frac{9}{16}$$

$$P(\text{tall and white}) = \frac{3}{4} \times \frac{1}{4} = \frac{3}{16}$$

$$P(\text{dwarf and coloured}) = \frac{1}{4} \times \frac{3}{4} = \frac{3}{16}$$

$$P(\text{dwarf and white}) = \frac{1}{4} \times \frac{1}{4} = \frac{1}{16}$$

This can be represented in tabular form:

	Coloured flowers	White flowers	
Tall plant	9/16	3/16	3/4
Dwarf plant	3/16	1/16	1/4
	3/4	1/4	

The **marginal probabilities** are shown at the end of each row and at the foot of each column. The probabilities in the body of the table may be obtained by multiplying together the corresponding row and column marginal probabilities.

We calculate the expected frequencies by multiplying each of these probabilities by the total frequency (160). The observed and expected frequencies are tabulated below:

	Coloured flowers		White flowers	
	Observed	Expected	Observed	Expected
Tall plant	85	90	32	30
Dwarf plant	34	30	9	10

Therefore,

$$Q = \frac{(85-90)^2}{90} + \frac{(32-30)^2}{30} + \frac{(34-30)^2}{30} + \frac{(9-10)^2}{10} = 1.0444$$

The 95th percentile of the χ^2 distribution with three degrees of freedom is 7.815, so we conclude that there is no evidence to disprove Mendel's theory at the 5% level. Note that we cannot say that we have *proved* Mendel's theory by this calculation.

Contingency tables

In Example 18, we obtained the probabilities from theoretical considerations. We can use the χ^2 goodness-of-fit test to test the hypothesis of statistical independence of two variables without having to assume any particular model for their joint distribution. In such cases we must *estimate* the probabilities from the table of observed frequencies, called a **contingency table**. (Remember that every parameter estimated reduces the number of degrees of freedom by one.)

Contingency means dependence, so a contingency table is simply a table that displays how two or more characteristics depend on each other.

Table 5.6 shows an example of a contingency table, where Variable 1 has three categories and Variable 2 has four categories. The frequencies are N_{ij}, the row and column totals are R_i and C_j, and the total frequency is T, so that

$$R_i = \sum_{j=1}^{4} N_{ij} \quad C_j = \sum_{i=1}^{3} N_{ij} \quad T = \sum_{i=1}^{3} R_i = \sum_{j=1}^{4} C_j$$

Table 5.6 A contingency table with three rows and four columns.

		Variable 2				
		A	B	C	D	
	I	N_{11}	N_{12}	N_{13}	N_{14}	R_1
Variable 1	II	N_{21}	N_{22}	N_{23}	N_{24}	R_2
	III	N_{31}	N_{32}	N_{33}	N_{34}	R_3
		C_1	C_2	C_3	C_4	T

The probability that an observation is in row i (the marginal probability for row i) may be estimated by R_i/T and the probability that it is in column j (the marginal probability for column j) by C_j/T and so, under the null hypothesis that the effects are independent, the probability associated with cell (i,j) is estimated by

$$p_{ij} = \frac{R_i}{T} \times \frac{C_j}{T} = \frac{R_i \times C_j}{T^2}$$

Hence the expected number for cell (i,j), under the hypothesis of independence, is estimated by

$$E_{ij} = Tp_{ij} = \frac{R_i \times C_j}{T}$$

For r rows and c columns, the number of degrees of freedom is given by $(r-1)(c-1)$. This is because there are rc frequencies, r row constraints, c column constraints (but this counts one overall constraint twice in the total $r+c$), leading to

$$rc - r - c + 1 = (r-1)(c-1)$$

Example 19

Every pupil in year 7 of a secondary school is given an English examination and a mathematics examination, each of which results in 'Pass', 'Fail' or 'Resit'. Table 5.7 is a contingency table of the results.

Table 5.7 Contingency table of examination results.

	English			
Mathematics	Fail	Resit	Pass	Total
Fail	20	15	11	46
Resit	13	30	17	60
Pass	12	22	55	89
Total	45	67	83	195

Investigate whether the two criteria are independent. In other words, are the results of the English examination related to the results of the mathematics examination? Or, put another way, if we know a pupil's English result, does that give us any information about what his or her mathematics result is likely to be?

Under the null hypothesis that the effects are independent, the expected frequencies are estimated from the formula

$$E_{ij} = \frac{R_i \times C_j}{T}$$

e.g.

$$E_{11} = \frac{46 \times 45}{195} = 10.615$$

$$E_{12} = \frac{46 \times 67}{195} = 15.805$$

and so on.

The resulting table of expected frequencies is

	English		
Mathematics	Fail	Resit	Pass
Fail	10.615	15.805	19.579
Resit	13.846	20.615	25.538
Pass	20.538	30.579	37.882

The χ^2 statistic is

$$Q = \frac{(20 - 10.615)^2}{10.615} + \frac{(15 - 15.805)^2}{15.805} + \cdots$$
$$= 8.297 + 0.041 + 3.759 + 0.052 + 4.272 + 2.855 + 3.550 + 2.407 + 7.735$$
$$= 32.967$$

Under H_0, Q will have a χ^2 distribution with $(r-1)(c-1) = 2 \times 2 = 4$ degrees of freedom. The significance probability of Q is 0.0000012, so H_0 is rejected without reservation – the English and Mathematics marks are definitely related.

Using Minitab:

```
MTB > chisquare analysis of table in c1 c2 c3

Expected counts are printed below observed counts

                C1         C2         C3       Total
      1         20         15         11         46
              10.62      15.81      19.58

      2         13         30         17         60
              13.85      20.62      25.54

      3         12         22         55         89
              20.54      30.58      37.88

    Total       45         67         83        195

    ChiSq =   8.297  +   0.041  +   3.759  +
              0.052  +   4.272  +   2.855  +
              3.550  +   2.407  +   7.735  =  32.967
    df = 4
```

EXERCISES ON 5.12

2. The distribution of five plant species, A, B, C, D, E, is under investigation at three different locations, I, II, III. The contingency table for the number of occurrences of each plant observed at each site is shown below:

		Species					Total
		A	B	C	D	E	
	I	10	22	38	8	66	144
Location	II	27	62	120	30	200	439
	III	45	100	207	49	342	743
	Total	82	184	365	87	608	1326

Is the relative abundance of each species the same at all locations?

3. In an eyesight examination a group of children was divided into two categories, those who wore spectacles (A), and those who did not (B). As a result of the test, their vision was classed as good, fair or bad. The children wearing spectacles were tested with and without them. The results are given overleaf:

	A with spectacles		A without spectacles		B	
	Boys	Girls	Boys	Girls	Boys	Girls
Good	145	182	79	69	5349	5151
Fair	307	263	240	235	1870	1989
Bad	67	48	207	212	602	625
Total	519	493	526	516	7821	7765

What conclusions would you draw, regarding (a) eyesight differences between the sexes, (b) whether it is useful to wear spectacles?

4. First-year university students chose a subsidiary subject for which they did not require an A-level qualification. The course thus assumed that no one had an A-level in the subject although several of them had. Out of 79 students who took the course, 46 did not have the A-level and the remaining 33 did. The grades on the course are illustrated in the table below. Carry out a χ^2 test reporting your conclusions clearly.

Grade:	A	B	C	D	E
A-level	4	7	10	8	4
Non-A-level	5	16	8	13	4

Summary

Single-sample tests

If $X \sim N(\mu, \sigma^2)$ and σ^2 is known, then

$$\frac{\bar{X} - \mu}{\sigma/\sqrt{n}} \sim N(0, 1)$$

If $X \sim N(\mu, \sigma^2)$ and σ^2 is unknown, then

$$t = \frac{\bar{X} - \mu}{s/\sqrt{n}}$$

has a t distribution with $n - 1$ degrees of freedom.

Matched pairs

Apply the single-sample t-test to the *differences*, $D_i = Y_i - X_i$.

Difference of means

If $X \sim N(\mu_x, \sigma^2)$ and $Y \sim N(\mu_y, \sigma^2)$ and σ^2 is known, then for m values of X and n values of Y,

$$\bar{X} - \bar{Y} \sim N\left(\mu_x - \mu_y, \sigma^2\left[\frac{1}{m} + \frac{1}{n}\right]\right)$$

If σ^2 is unknown, then

$$t = \frac{\bar{X} - \bar{Y}}{s\sqrt{1/m + 1/n}} \quad \text{where} \quad s^2 = \frac{(m-1)s_x^2 + (n-1)s_y^2}{m+n-2}$$

has a t distribution with $m + n - 2$ degrees of freedom.

Wilcoxon's signed-ranks test

Calculate the *differences* and rank them in order; then T^+ is the sum of the ranks corresponding to positive differences.

For ranks r_i,

$$E(T^+) = \frac{1}{2}\sum_{i=1}^n r_i \quad \text{and} \quad Var(T^+) = \frac{1}{4}\sum_{i=1}^n r_i^2$$

The Mann–Whitney test

Given two samples, $X_1, X_2, \ldots, X_m$ and $Y_1, Y_2, \ldots, Y_n$, pool the two samples and then rank them. Then calculate S, the sum of the ranks of the X data.

$$E(S) = m\bar{r} \quad \text{and} \quad Var(S) = ms_r^2\left(1 - \frac{m}{m+n}\right)$$

Tests for proportions

Single sample: if p is the true proportion,

$$\frac{\hat{p} - p}{\sqrt{p(1-p)/n}} \sim N(0, 1)$$

Two sample: under H_0: $p_1 = p_2$,

$$\hat{p}_1 - \hat{p}_2 \sim N\left(0, p(1-p)\left(\frac{1}{n_1} + \frac{1}{n_2}\right)\right)$$

where p is the true (common) proportion, which we can estimate using

$$\hat{p} = \frac{n_1\hat{p}_1 + n_2\hat{p}_2}{n_1 + n_2}$$

Variance-ratio test

Given two samples, $X_1, X_2, \ldots, X_m$ and $Y_1, Y_2, \ldots, Y_n$, with equal variances (i.e. $\sigma_x^2 = \sigma_y^2$),

$$\frac{s_x^2}{s_y^2} \sim F(m-1, n-1)$$

Regression coefficients

$$\hat{\beta}_0 \sim N\left(\beta_0, \sigma^2\left[\frac{1}{n} + \frac{\bar{x}^2}{S_{xx}}\right]\right) \qquad \hat{\beta}_1 \sim N\left(\beta_1, \frac{\sigma^2}{S_{xx}}\right)$$

$$\hat{\beta}_0 + \hat{\beta}_1 x \sim N\left(\beta_0 + \beta_1 x, \sigma^2\left[\frac{1}{n} + \frac{(x-\bar{x})^2}{S_{xx}}\right]\right)$$

$$\text{Fresh observation} \sim N\left(\beta_0 + \beta_1 x, \sigma^2\left[1 + \frac{1}{n} + \frac{(x-\bar{x})^2}{S_{xx}}\right]\right)$$

Correlation coefficient

$$r\sqrt{\frac{n-2}{1-r^2}}$$

has a t distribution with $n-2$ degrees of freedom.

Chi-squared goodness of fit test

$$Q = \sum_{i=1}^{k} \frac{(O_i - E_i)^2}{E_i}$$

has (approximately) a χ^2 distribution with n degrees of freedom, given by

$$n = k - con - par$$

where

k = number of observed frequencies
con = number of constraints on the expected frequencies
par = number of parameters estimated from the data in order to calculate the expected frequencies.

Contingency tables

The expected number for cell (i, j) is

$$E_{ij} = \frac{R_i \times C_j}{T}$$

where R_i and C_j are the row and column totals, respectively, and T is the total frequency.

For r rows and c columns, the number of degrees of freedom is given by $(r-1)(c-1)$.

FURTHER EXERCISES

1. It is suggested that supermarkets in Swansea tend to charge more for their goods than those in Cardiff. A shopper in Cardiff arranges with a shopper in Swansea to purchase an agreed assortment of goods and compare prices. The two cities have 10 supermarket chains in common, which we shall call A, B, C,

..., J, and the shoppers visit each in turn in consecutive weeks, recording the following prices:

Store:	A	B	C	D	E	F	G	H	I	J
Swansea	12.08	12.81	12.74	13.54	14.86	14.68	12.64	15.23	13.83	12.64
Cardiff	11.62	11.69	12.57	13.32	13.15	14.04	11.76	13.63	12.95	12.59

Perform a suitable t-test, stating whether it is paired or unpaired, one-tailed or two-tailed, to test at the 5% significance level the theory that Swansea's supermarket prices are higher than Cardiff's.

2. A well-known supermarket chain markets its own brand of dishwasher powder. Its market research suggests that most of its customers are willing to pay more for a product that is more environmentally friendly. Producing such a product incurs a 5% increase in production costs from 40p per packet to 42p per packet. The company adopts the policy of transferring its whole production to the new product, and increasing the price from £1 to £1.10. The average number of packets sold (1000s) per week in the final six months of retailing the old product and the first six months of the new product are shown below for shops in 15 different cities.

Recalling the central limit theorem, is it reasonable to assume that these values will be normally distributed? If so, use a suitable t-test to determine whether the change in the product has resulted in a significant increase in profits. Otherwise use a suitable non-parametric test.

City	Old product	New product
London	50	45
Glasgow	60	55
Manchester	51	49
Cardiff	59	51
Swansea	57	54
Leeds	59	57
Newcastle	54	49
Birmingham	55	47
Liverpool	54	59
Aberdeen	61	52
Bristol	56	51
Bradford	55	54
Exeter	51	50
Sheffield	59	57
Norwich	60	53

3. Using the cholesterol data (Dataset 1, Appendix A), perform a suitable hypothesis test to test the following theories:
   (a) that the male patients are significantly taller than the female patients;
   (b) that the weight loss experienced depends on whether the patient was seen by a health visitor or by a dietician.

4. Returning to Section 4.4, Example 1, the data quoted there was collected prior to the start of an opencast mine near the village. Eighteen weeks after the start of operations, the equivalent weekly figures were

$$6\ 7\ 11\ 10\ 11\ 5\ 4\ 12\ 8\ 4\ 9\ 14\ 7\ 4\ 16\ 12\ 11\ 4$$

Perform an appropriate two-sample test to highlight whether there has been a significant change in weekly episode rate.

5. Show directly that

$$Var[\bar{y} + \hat{\beta}_1(x - \bar{x})] = \sigma^2 \left(\frac{1}{n} + \frac{(x - \bar{x})^2}{S_{xx}} \right)$$

by expressing $\bar{y} + \hat{\beta}_1(x - \bar{x})$ in the form $\sum a_i y_i$ and computing $\sum a_i^2$.

6. With the shove-halfpenny data (see Section 4.7 of Chapter 4), test the hypotheses that the intercept equals zero and the slope equals one.

7. With the data you collected from Tutorial Problem 1 in Chapter 3 (the dart data), test the hypothesis that the population mean x value equals zero, and also that the population mean y value equals zero (corresponding to being on target).

8. Also with the dart data, test the hypothesis that the population variance of the x values equals the population variance of the y values.

9. Again with the dart data, your answers to Exercises 7 and 8 above depend on the assumption that the x and y data are normally distributed (do they?). Test this hypothesis by calculating expected frequencies to match appropriate observed frequencies and using the χ^2 goodness-of-fit test.

10. Finally, with the dart data, look at the sample distribution of the r^2 data (square of distance from the target value). Use the χ^2 goodness-of-fit test to test the hypothesis that the appropriate distribution to model it is the exponential distribution. You will need to estimate the mean parameter of this distribution, and again calculate expected frequencies corresponding to observed frequencies which you have carefully chosen.

6 • More Advanced Statistics

The models and techniques in statistics presented here are just part of the foundation of modern statistics. In practical examples of statistical analysis it is only too easy to ask a simple question which immediately takes you beyond the remit of this book. These comments help to map out the structure of more advanced statistical techniques that are available.

There are some very obvious extensions to some of the procedures included in this book. The two-sample t-test addresses the question of testing the equality of two population means, and it is natural to ask – why stop at two? The techniques in **analysis of variance** address this. They have long been the pride and joy of agricultural scientists and, more recently, experimental psychologists and quality control engineers. Similarly, the technique of linear regression can be extended to multiple regression where we require to predict one variable given a whole set of predictor variables, some of which may be redundant.

All such modelling tends to assume that individual observations are statistically independent. When observations occur through time (**time series**), such as temperature, rainfall, stock exchange yields, observations near in time are going to be correlated, and the models constructed must reflect this. Similarly, when observations are not in one dimension (e.g. the wear at different places across the width of a car tyre), correlations between values in different dimensions must be assumed at least initially, and this points to the vast subject of **multivariate statistical analysis**.

In the majority of the statistical procedures we looked at in this book, and in the extensions mentioned above, the data is usually assumed to be normally distributed. If, on the other hand, we were studying the time to failure of a physical component (reliability engineering) or a human machine (medical statistics) we cannot make this assumption (an exponential distribution would be more appropriate) and so the whole methodology of regression modelling has to be extended to a different distribution. In the last 30 years, one of the great triumphs of mathematical statistics has been to develop the basic statistical procedures of analysis of variance and multiple regression into a unified framework which allows a wide range of distributional assumptions to be made – the subject area is called **generalised linear models**.

No description of the field of modern statistics would be complete without mentioning Bayesian statistical analysis, which is becoming more and more popular as modern computational facilities make it feasible to perform the appropriate analyses. The basic idea is to allow the users of statistics to express their a priori information concerning, say, a parameter, by means of a probability distribution interpreted in terms of degrees of belief. The data analysis cycle then shows how this belief is updated in the presence of the data.

Statisticians have always been at the forefront in using newly acquired computational facilities. Another area where this has been exploited successfully is in simulation. In the search for more and more realistic statistical models, the bounds on the mathematical ingenuity to solve the statistical problems mathematically are soon exceeded. In such a situation, the computer can be used to simulate the model, so that appropriate methods of analysis can be investigated and evaluated empirically, much as an experimental scientist in chemistry or biology would do. Graphical tools on modern computers are so good too that there has been in recent years some very interesting mathematics done to show how these tools can best be used, say, within the topic of regression.

Clearly the list is endless, and there is unfortunately a huge lag between the development of appropriate statistical models and methods, and their routine application in situations to which they apply. There will always be a need for statisticians, both theoretical and applied, to propose and evaluate new methods in statistics. Equally so there will always be a need to make sure that the information and data collected by scientists, engineers, countries, is of sound quality so that the application of the correct statistical technique and the interpretation of the consequential results lead to sensible judgements and decisions that do and will continue to affect the lives of all of us!

Appendix A: Datasets

Dataset 1: Cholesterol data

This data was collected by a general practice which conducted an interventionist campaign targeting people at risk from heart-related problems. They were seen by an advisor (Health Visitor or Dietician) and encouraged to adopt a 'better' diet. The variables are:

who	HV = Health visitor D = Dietician
bmi	Body Mass Index (Weight divided by Height squared)
wtb	Weight before (kg)
wta	Weight after (kg)
ht	Height (m)
cholb	Cholesterol level before the diet (mmol/l)
chola	Cholesterol level after the diet
sex	M = Male F = Female
age	Age (years)

who	bmi	wtb	wta	ht	cholb	chola	sex	age
HV	26	76	80	1.68	7.58	6.31	F	49
HV	25	70	70	1.66	7.53	6.71	F	40
HV	28	72	66	1.58	7.22	6.87	F	51
HV	28	80.5	81	1.67	7.36	7.46	F	53
HV	22	57.5	57	1.59	6.24	6.32	F	31
HV	29	82	82	1.68	7.57	8.70	F	54
D	31	95	93	1.73	7.04	6.44	M	52
D	28	80	80	1.68	7.11	7.43	M	38
HV	22	59	59	1.61	7.04	6.75	F	53
D	34	107	99	1.77	6.76	5.41	M	38
D	23	66	67	1.67	6.13	4.77	M	42
D	29	75.5	76	1.59	7.66	7.43	F	48
D	24	99	99	2.00	7.24	7.04	M	39
HV	22	55.5	56	1.56	6.49	6.58	F	40
D	29	73.5	74	1.58	7.29	6.38	F	37
HV	29	73.5	72	1.59	7.20	7.38	F	53
D	24	60	60	1.58	7.20	7.04	F	54
D	17	53	53	1.73	7.43	6.66	F	43
HV	19	47	47	1.56	7.45	6.45	F	54
D	26	62	62	1.53	7.38	7.17	F	39
D	27	86	86	1.78	7.41	7.16	F	39
D	30	94	94	1.77	7.55	7.63	F	48
D	23	62.5	63	1.62	7.58	6.02	F	41
D	27	100	100	1.92	7.15	7.22	M	48

who	bmi	wtb	wta	ht	cholb	chola	sex	age
HV	22	72.5	73	1.79	7.08	6.14	M	35
D	28	91	89	1.80	7.15	5.70	M	39
D	27	74.5	70	1.65	7.50	6.92	M	37
HV	22	72	72	1.78	6.58	5.49	M	42
D	23	54	51	1.53	7.99	7.27	F	56
HV	26	72.4	72	1.65	7.03	7.23	M	51
HV	25	82	76	1.80	6.86	5.33	M	40
D	22	68	68	1.74	6.81	6.65	M	47
HV	26	85	86	1.78	7.16	7.40	M	55
D	25	76.5	76	1.74	6.78	6.35	M	44
HV	31	98.5	98	1.76	6.76	8.19	M	39
D	24	74.5	75	1.73	6.48	5.99	M	45
D	25	78	78	1.74	7.51	7.57	M	40
D	26	80.5	81	1.75	7.34	7.13	M	55
D	27	85	85	1.75	7.35	6.85	M	47
HV	28	91.5	92	1.80	7.29	7.05	M	36
D	25	63.2	63	1.58	7.52	6.25	F	42
HV	25	69	69	1.63	7.19	7.06	F	55
HV	24	69	69	1.67	7.31	6.81	F	54
D	22	57.5	54	1.61	6.89	6.33	F	26
D	22	68	68	1.73	7.32	6.88	M	47
HV	23	65	61	1.67	7.02	6.94	F	54
HV	47	117	115	1.57	7.15	5.60	F	52
D	35	93.2	93	1.63	6.84	7.20	F	39
D	25	80.5	73	1.78	6.87	6.40	M	42
D	28	91	75	1.79	7.91	5.43	M	44
D	25	66.3	66	1.60	6.53	7.01	F	38
D	35	79	76	1.49	6.69	7.43	F	49
D	23	62	59	1.67	7.63	5.37	F	43
D	23	76	70	1.79	6.80	6.24	M	38
D	23	62	53	1.63	7.44	5.58	F	39
HV	32	73.5	71	1.50	5.25	5.45	F	44
D	24	71.5	69	1.72	7.58	6.88	M	38
D	36	97.2	97	1.63	5.62	5.21	F	42
HV	27	63.5	60	1.52	6.93	6.93	F	39
HV	32	92	86	1.68	6.69	7.01	M	37
HV	26	84.25	83	1.80	7.02	7.46	M	43
D	27	75.6	74	1.66	7.34	8.33	M	46
HV	19	59	58	1.74	7.01	5.49	M	36
D	28	82.5	75	1.71	7.00	6.16	M	29
HV	24	86	84	1.88	7.81	6.22	M	48
D	25	84.5	86	1.82	7.62	7.16	M	33
D	25	78	73	1.76	6.55	5.17	M	33
D	21	55	51	1.60	6.25	5.38	F	37
D	24	73.5	61	1.75	6.93	6.37	M	38
D	23	67.3	67	1.71	7.46	7.98	M	37
HV	28	82	82	1.71	7.33	7.03	M	35

who	bmi	wtb	wta	ht	cholb	chola	sex	age
HV	25	66.5	68	1.61	7.70	6.99	F	51
HV	23	66	70	1.67	7.28	6.27	F	44
HV	22	56	53	1.59	7.61	5.43	F	38
HV	34	124	116	1.90	7.97	8.62	M	31
D	30	87	85	1.69	6.68	5.36	F	58
HV	22	68.5	67	1.73	6.99	6.54	M	46
HV	24	56.5	52	1.53	7.06	7.39	F	48
HV	24	70.25	68	1.68	6.92	6.81	M	54
HV	26	77.5	73	1.70	6.71	8.99	M	55
HV	29	87.5	88	1.71	7.75	7.36	M	52
HV	25	76.5	69	1.73	6.92	6.70	M	53
HV	26	85	84	1.79	6.09	5.73	M	34
HV	30	80.5	76	1.62	7.92	5.62	F	46
D	28	71.5	72	1.58	7.08	6.97	F	44
HV	24	60.4	60	1.56	7.46	7.85	F	34
D	37	91.6	88	1.56	7.07	7.06	F	53
HV	27	71	69	1.61	6.81	6.86	F	46
D	23	55	53	1.52	7.56	6.14	F	48
HV	18	54.5	54	1.74	7.46	6.17	M	54
D	24	66.5	67	1.65	7.71	5.78	F	51
HV	27	74.5	76	1.65	7.16	8.04	M	37
HV	24	74	74	1.75	6.70	5.06	M	38
D	40	98	82	1.56	6.50	6.58	F	44
D	36	87.5	87	1.55	5.90	7.07	F	42
D	26	67	65	1.58	8.16	6.57	F	55
HV	35	94	94	1.62	4.43	3.93	F	32
D	23	56.4	51	1.55	7.67	6.08	F	24
D	27	79	79	1.69	7.19	7.63	M	40
D	26	84	81	1.78	8.29	6.78	M	38
HV	25	73.4	72	1.70	9.29	7.34	M	57
HV	32	102	96	1.78	9.55	8.08	M	43
D	22	73.5	74	1.74	8.32	6.21	M	36
D	26	68.8	65	1.60	8.67	7.43	F	47
D	30	88.3	88	1.70	8.82	7.69	M	51
D	30	71.75	66	1.54	8.97	6.45	F	35
HV	29	81.75	83	1.66	8.20	7.78	M	39
HV	33	89.6	81	1.64	9.73	7.38	F	49
HV	29	76	72	1.60	8.28	7.48	F	53
D	23	53.5	50	1.51	8.76	7.88	F	50
HV	20	59	57	1.69	9.60	7.71	M	38
HV	28	98	98	1.86	8.48	6.84	M	34
D	19	67.65	68	1.85	8.14	8.28	M	46
HV	28	77.5	78	1.65	8.13	6.89	F	34
HV	27	87	87	1.78	8.58	6.22	M	55
HV	34	92	92	1.63	8.52	8.18	F	27
D	21	52	52	1.56	8.89	4.51	F	50
HV	28	75.25	75	1.62	8.13	7.83	F	55

Dataset 2: Olympic relay data

This data lists the winning times in seconds for the 4 × 100 m relay race at the Olympic Games from 1928 to 1988 for both men and women.

Year	Men			Women		
	Gold	Silver	Bronze	Gold	Silver	Bronze
1928	41.00	41.20	41.80	48.40	48.80	49.00
1932	40.00	40.90	41.20	47.00	47.00	47.60
1936	39.80	41.10	41.20	46.90	47.60	47.80
1948	40.60	41.30	41.50	47.50	47.60	48.00
1952	40.10	40.30	40.50	45.90	45.90	46.20
1956	39.50	39.80	40.30	44.50	44.70	44.90
1960	39.50	40.10	40.20	44.50	44.80	45.00
1964	39.00	39.30	39.30	43.60	43.90	44.00
1968	38.20	38.30	38.40	42.80	43.30	43.40
1972	38.19	38.50	38.79	42.81	42.95	43.36
1976	38.33	38.66	38.78	42.55	42.59	43.09
1980	38.26	38.33	38.53	41.60	42.10	42.43
1984	37.83	38.62	38.70	41.65	42.77	43.11
1988	38.19	38.28	38.40	41.98	42.09	42.75

Dataset 3: Population of Ireland in 1841

Age	Male	Female
0–4	523727	505798
5–9	544854	531351
10–14	518876	499473
15–19	432037	453723
20–24	372831	413012
25–29	293497	318170
30–34	269955	302482
35–39	185123	194874
40–44	231942	252037
45–49	132456	135296
50–54	167140	180901
55–59	92306	89335
60–64	119519	138203
65–69	44854	42684
70–74	44587	47431
75–79	18736	17178
80–84	17074	21903
85–	7022	8220

Dataset 4: English football league goals for 18 December 1994

All goals scored are listed (home goals first) and their times. If a match number is missing, then no goals were scored in that match.

Premier Division

Match number	1	1	1	3	6	6	6	7	7	8	8	8
Home/away	H	A	A	A	H	A	A	A	A	H	H	H
Goal time	85.2	23.8	84.5	46.6	67.3	34.1	61.7	59.7	84.0	6.4	9.6	56.9

Division 1

Match number	1	1	1	1	1	1	2	2	3	4
Home/away	H	H	H	A	A	A	H	H	H	A
Goal time	13.1	57.1	87.6	4.7	7.1	89.4	57.1	75.5	67.7	59.6

Match number	5	5	5	5	6	7	7	8
Home/away	H	H	A	A	A	H	H	A
Goal time	11.0	57.9	24.6	49.1	9.7	53.5	74.8	81.0

Division 2

Match number	1	1	2	2	3	3	3	3	3	3	3	4	4
Home/away	H	H	H	A	H	H	H	H	H	H	H	H	A
Goal time	1.8	86.3	1.1	70.8	2.4	7.7	47.8	56.0	49.4	73.3	87.0	58.8	2.8

Match number	5	5	5	5	6	6	7	7	8	8	8	8
Home/away	H	H	H	A	H	H	H	A	H	H	H	A
Goal time	55.7	71.1	75.7	37.9	89.3	81.5	62.8	71.5	16.3	27.3	87.4	42.2

Division 3

Match number	1	1	2	2	2	2	3	4	4	4	5	5	5
Home/away	A	A	H	H	H	A	H	H	H	A	H	A	A
Goal time	6.5	51.9	27.0	44.1	70.0	48.6	87.6	24.4	31.0	63.5	23.3	27.1	49.0

Match number	6	6	6	7	7	7	7	7	7	7	8	8
Home/away	H	H	A	H	H	H	H	A	A	A	H	A
Goal time	41.1	87.6	58.4	26.3	47.0	54.6	57.5	36.0	42.1	88.8	3.8	80.9

Appendix B: Sampling from a Finite Population

Sample surveys are concerned with sampling from a **finite** population. The details of this are (with one exception – the two-sample Mann–Whitney test) beyond the scope of this elementary text. Thus the essential results needed for the non-parametric procedure have been relegated to this appendix. The casual reader need only understand the basic conclusions concerning the mean and variance of the sample mean under sampling with and without replacement and why they differ.

We start by defining the population of interest – it has N objects, and the attribute under investigation has measurements $x_1, x_2, \ldots, x_N$. The population mean μ and the population variance σ^2 are given by the expressions

$$\mu = \frac{(x_1 + x_2 + \cdots + x_N)}{N} \quad \text{and} \quad \sigma^2 = \frac{\sum_{i=1}^{N}(x_i - \mu)^2}{N}$$

Note the divisor N here. It will be also useful to use a divisor $N - 1$, so we will define

$$S^2 = \frac{\sum_{i=1}^{N}(x_i - \mu)^2}{N - 1}$$

Sampling with replacement

Suppose the random variables $X_1, X_2, \ldots, X_n$ are n values chosen **with replacement** from the population at random. This means that for any X_i the probability that it takes any particular one of the N values in the population is $1/N$. As a consequence, it is easy to show that

$$E(X_i) = \mu \quad Var(X_i) = \sigma^2$$

Hence the sample mean $\bar{X} = (X_1 + X_2 + \cdots + X_n)/n$ has mean and variance

$$E(\bar{X}) = \mu \quad Var(\bar{X}) = \frac{\sigma^2}{n}$$

(because the X_i are statistically independent).

Sampling without replacement

When we come to consider sampling **without replacement** from first principles, it is not so clear that all the X_i now have the same distribution – but they do. Consider just X_1 and X_2 to demonstrate that they do. Clearly $P(X_1 = x_i) = 1/N$ as before as there are N choices for the first of the sample. For X_2, the argument is more complicated as we need to use conditional probability:

$$P(X_2 = x_i) = P(X_2 = x_i \mid X_1 \neq x_i) P(X_1 \neq x_i)$$

(as we sample without replacement, it is not possible for X_2 to equal x_i if X_1 equals x_i). Hence

$$P(X_2 = x_i) = \frac{1}{N-1}\left(1 - \frac{1}{N}\right) = \frac{1}{N}$$

and so the distribution of X_2 (and the others) is the same as if we had sampled with replacement. Hence it follows that $E(\bar{X}) = \mu$ as before. The difference comes when we consider the variance of $\bar{X}$. If we consider the extreme case when $n = N$, i.e. we have sampled the whole population, then we **know** the population mean μ **exactly** and so the variance of $\bar{X}$ is zero. Intuitively then, we would expect the variance of $\bar{X}$ to decrease as the sample size n increases. The appropriate expression for this variance is derived below – the details can be skipped by the faint-hearted!

To make the algebra as simple as possible, and without loss of generality, let us assume that μ is zero. (This can always be done simply by defining a new population where all the x_i had μ subtracted from them – this will not affect variance calculations at all.) This means that $x_1 + x_2 + \cdots + x_N = 0$ and $\sigma^2 = Var(X_i) = E(X_i^2)$.

To calculate the variance of $\bar{X}$, equal to $E[(X_1 + X_2 + \cdots + X_n)^2/n^2]$, we can expand the squared bracket to give n squared terms X_i^2 with expected value σ^2 and $n(n-1)$ cross-product terms $X_i X_j$ whose expected value needs to be evaluated (because X_i and X_j are now **dependent random variables**).

Consider X_1 and X_2 as before:

$$E(X_1 X_2) = \frac{\sum_{i=1}^{N}\sum_{j\neq i} x_i x_j}{N(N-1)} = \frac{\sum_{i=1}^{N} x_i \sum_{j\neq i} x_j}{N(N-1)} = \frac{\sum_{i=1}^{N} x_i(0 - x_i)}{N(N-1)}$$

$$= \frac{-\sum_{i=1}^{N} x_i^2}{N(N-1)} = -\frac{\sigma^2}{N-1}$$

Since there are n squared terms all having the same expected values and $n(n-1)$ cross-product terms whose expected values are again equal by symmetry, we have

$$Var(\bar{X}) = \frac{1}{n^2} E(X_1 + X_2 + \cdots + X_n)^2 = \frac{1}{n^2}\left(n\sigma^2 - n(n-1)\frac{\sigma^2}{N-1}\right)$$

$$= \frac{\sigma^2}{n}\left(\frac{N}{N-1}\right)\left(1 - \frac{n}{N}\right)$$

Finally, bearing in mind the definition of S^2, we have

$$Var(\bar{X}) = \frac{S^2}{n}\left(1 - \frac{n}{N}\right)$$

This formula is now very easy to interpret: think of S^2 as the population variance, rather than as originally defined. The variance of $\bar{X}$ has to be reduced by dividing by the sample size n (as in sampling with replacement). In addition, however, it has to be reduced by multiplying by the **sample fraction** $(1 - n/N)$ which represents the proportion of the population not sampled – if this is small the variance of $\bar{X}$ will be correspondingly smaller than if the sampling fraction were approximately one.

Solutions

Chapter 1

Ex.2 p.4 $\binom{5}{2} = 10$ selections. $3 \times 2 = 6$ result in $1-1$.

Ex.1 p.6 (a) Ordinal, (b) Nominal, (c) Ordinal, (d) Ratio, (e) Ordinal, (f) Ordinal.

Ex.2 p.6 (a) and (c).

Chapter 2

Ex.1 p.16 Round the data to the nearest integer:

```
   2    9  78
   7   10  16678
  24   11  00012246677788889
 (19)  12  0111112223444566888
  27   13  00133477889
  16   14  345689
  10   15  04456
   5   16  038
   2   17  8
   1   18  6
```

Ex.2 p.25 $i_1 = 10 \times 0.25 + 0.5 = 3$, $i_2 = 10 \times 0.5 + 0.5 = 5.5$, $i_3 = 10 \times 0.75 + 0.5 = 8.$
Dataset A: 4.5 4.7 6.0, Dataset B: 3.0 4.5 7.0
Dataset C: 1.4 5.1 8.1, Dataset C: 3.0 3.25 4.0

Ex.1 p.35 (a) -0.035, (b) -0.549, (c) 0.962.

Ex.1 p.38 $\bar{x} = 4.4$, $s = 1.577$.

```
   1    1  9
   8    2  0133499
  13    3  13469
 (13)   4  0345566788899
  10    5  23336
   5    6  356
   2    7
   2    8  13
```

Chapter 3

Ex.1 p.42 $P(10) = \dfrac{1^2 \times \pi}{8^2 \times \pi} = 0.015625, \quad P(5) = \dfrac{3^2 \times \pi}{8^2 \times \pi} - P(10) = 0.125,$

$P(3) = \dfrac{5^2 \times \pi}{8^2 \times \pi} - P(5) - P(10) = 0.25,$

$P(0) = 1 - P(3) - P(5) - P(10) = 0.609375,$
$E(X) = 3 \times 0.25 + 5 \times 0.125 + 10 \times 0.015625 = 1.53125.$

Ex.1 p.47 $P(\text{Resistor faulty}) = \dfrac{95}{1000} = 0.095 \ (estimate).$

Number faulty in each box is $Bi(10, 0.095)$.
$P(\text{Box rejected}) = P(X \geq 3)$ for $X \sim Bi(10, 0.095) = 0.062.$
Expected number of rejections $= 100 \times 0.062 = 6.2.$

Ex.2 p.47 $P(\text{Correct answer}) = 0.25.$
Number of correct answers is $Bi(100, 0.25)$.
$P(\text{Pass}) = P(X \geq 35)$ for $X \sim Bi(100, 0.25) = 0.016.$

Ex.3 p.47 $P(\text{Breaks pane}) = \dfrac{1}{11}.$

Let $X = $ Number of panes broken in 12 trials. $X \sim Bi\left(12, \dfrac{1}{11}\right).$

$P(\text{Cuts 10 correctly}) = P(X \leq 2)$ for $X \sim Bi\left(12, \dfrac{1}{11}\right) = 0.911.$

Ex.4 p.48 $P(\text{Component fails}) = 0.38.$ Suppose n components are taken.
Let X be the number of components that fail. $X \sim Bi(n, 0.38)$.
We require $P(X \geq 2) \geq 0.99$.

n	8	9	10	11	12	13	14	15
$P(X \geq 2)$	0.871	0.912	0.940	0.960	0.973	0.982	0.988	0.992

Ex.1 p.53 (a) 2, (b) 1 or 2, (c) 1.782, (d) $2 - 1.782 = 0.218,$

(e) x	0	1	2	3	4	5	6
$P(X = x)$	0.135	0.271	0.271	0.180	0.090	0.036	0.012
$P(X \geq x)$	0.135	0.406	0.677	0.857	0.947	0.983	0.995

Must cater for three **more** tankers daily.

Ex.2 p.53 There are $65 \times 55 = 3575$ characters per page.
$E(X) = 3575 \times 0.005 = 17.875$, so
$X \sim Po(17.875)$. $P(X < 4) = 0.00002.$
For $p = 0.001$, $E(X) = 3575 \times 0.001 = 3.575$ and $P(X \leq 3) = 0.52.$

Ex.3 p.53 Let X be the demand and let Y be the number of books on loan.
(a) $P(Y = 5) = P(X \geq 5) = 1 - P(X < 5) = 1 - 0.629 = 0.371.$

(b)

y	0	1	2	3	4	5
$P(Y=y)$	0.0183	0.0733	0.1465	0.1954	0.1954	0.371

$E(Y) = 0 \times 0.0183 + 1 \times 0.0733 + \cdots + 5 \times 0.371 = 3.59$.

(c) Expected frequencies: multiply the probabilities in (b) by 72:

Copies	0	1	2	3	4	5
Expected frequency	1.3	5.3	10.5	14.1	14.1	26.7

These are fairly consistent with the observed frequencies.

Ex.1 p.55 $P(\text{prize}) = 1 - \dfrac{\binom{6}{0}\binom{54}{6} + \binom{6}{1}\binom{54}{5} + \binom{6}{2}\binom{54}{4}}{\binom{60}{6}} = 0.0103$.

$P(\text{jackpot}) = \dfrac{1}{\binom{60}{6}} = 1.997 \times 10^{-8}$.

Ex.1 p.56 Expected frequencies: 155.5, 93.7, 56.4, 34, 20.5, 12.3, 7.4, 4.5, 2.7, 1.6.

Ex.1 p.57 $P(\text{10th toss}) = 0.123$. $P(\text{before 10th toss}) = 0.5$.

Ex.1 p.60 (a) $a = \dfrac{2}{3}$, (b) $E(X) = \dfrac{4}{3}$, (c) $Var(X) = \dfrac{7}{18}$.

Ex.2 p.61 $f(x) = (b-a)^{-1}$ for $a \leq x \leq b$, $E(X) = \dfrac{a+b}{2}$, $Var(X) = \dfrac{(b-a)^2}{12}$.

Ex.1 p.63 $\hat{\mu} = 525.5$. Expected frequencies: 306.9, 118.5, 45.8, 17.7, 9.5, 1.7.

Ex.1 p.66 Percentage defective = 9.56%.
Maximum allowed standard deviation = 0.001520.

Ex.2 p.66 0.0062.

Ex.3 p.66 $\bar{x} = 2.5226$, $s = 1.52685$, $P(X > 5) = 0.052$.
Expected number > 5 is $50 \times 0.052 = 2.6$ (observed is 0).

Ex.4 p.69 Using the normal approximation: (a) 0.722, (b) 0.41, (c) 0.007.
(Actual binomial values are 0.722, 0.441, 0.00344.)

Ex.5 p.69 0.493.

Ex.6 p.74 (a) 0.5811, (b) 0.3625, (c) 0.0797.

Ex.7 p.77 $\bar{X} \sim N(20, 0.0025)$. $P(\bar{X} < 19.9) = 0.02275$.

Ex.8 p.77 Total $\sim N(37.5, 6.75)$. $P(\text{Total} > 40) = 0.168$. Expect 61.3 days in 365.

Ex.1 p.79 $A(p) = (1-p)^8(1 + 8p + 36p^2)$.

Ex.2 p.80 $P(\text{2-engine}) = p(2-p)$, $P(\text{4-engine}) = p^2(6 - 8p + 3p^2)$.

$p^2(6 - 8p + 3p^2) > p(2-p)$ provided $p > \dfrac{2}{3}$.

Solutions

Ex.3 p.80

Females	0	1	2	3	4
Observed	246	875	1250	789	183
Expected	236.7	888.7	1251.1	782.8	183.7

$p = \dfrac{6474}{13372} = 0.48415$. Sample variance $= 1{:}007$.

Fitted variance $= 0.999$.

Ex.4 p.80 $p = P(\text{rain} < 50) = 0.9938$. $P(10 \text{ years} < 50) = p^{10} = 0.94$.

Ex.5 p.80 2331 males, 4736 pigs. $p = \dfrac{2331}{4736} = 0.4922$.

Number of males	Size of litter			
	4	5	6	7
0	4	4	4	3
1	14	19	22	22
2	20	37	53	64
3	13	36	69	104
4	3	17	50	101
5	–	3	19	58
6	–	–	3	19
7	–	–	–	3

Ex.6 p.80 Let $p = P(\text{component fails})$. Profit $= 10(1-p) - 5p = 5(2-3p) = 5$ for $p = \dfrac{1}{3}$.

$p = 1 - e^{-x/100} = \dfrac{1}{3}$ for $x = -100 \log \dfrac{2}{3} = 40.55$.

Ex.7 p.81 49.375 pence per litre.

Chapter 4

Ex.1 p.84

Samples	1 2 3	1 2 4	1 3 4	2 3 4
Median	2	2	3	3

$P(\text{Median} = 2) = P(\text{Median} = 3) = 0.5$.

Ex.2 p.84 Distribution of Median:

Value	1	2	3	8
Probability	0.15625	0.34375	0.34375	0.15625

Expected value $= 3.125 \neq$ Population mean of 3.5.

Ex.3 p.84 The unbiased estimator based on the largest number of observations is better as its variance is smaller.

Ex.4 p.84

	Sample mean	Sample standard deviation
Observed proportion	0.404	0.1034
Reciprocal of average run length	0.4428	0.1149

First method is better because (a) it is unbiased and (b) it has smaller variance.

Ex.1 p.87 $P(T \geq 1) = 1 - (1-p)^2 = 0.025$ (or 0.05 for 90%) when $p = 1 - \sqrt{0.975}$ (or $1 - \sqrt{0.95}$).
$P(T \leq 1) = 1 - p^2 = 0.025$ (or 0.05) when $p = \sqrt{0.975}$ (or $\sqrt{0.95}$).

Ex.2 p.87 Let T be the number in the 100 sampled that have antibodies present.

$$P(T \geq 38) = P(T > 37.5) \approx P\left(\frac{T - 100p}{\sqrt{100p(1-p)}} > \frac{37.5 - 100p}{\sqrt{100p(1-p)}}\right)$$

$= 0.025$ when $\dfrac{37.5 - 100p}{\sqrt{100p(1-p)}} = 1.96$, i.e. $p = 0.2864$.

$$P(T \leq 38) = P(T < 38.5) \approx P\left(\frac{T - 100p}{\sqrt{100p(1-p)}} < \frac{38.5 - 100p}{\sqrt{100p(1-p)}}\right)$$

$= 0.025$ when $\dfrac{38.5 - 100p}{\sqrt{100p(1-p)}} = -1.96$, i.e. $p = 0.4829$.

Ex.3 p.88 95% confidence interval for p: $0.511 \pm 0.01 = (0.501, 0.521)$.

Ex.1 p.91 $l(p) = \log(p^r(1-p)^{n-r}) = r \log p + (n-r)\log(1-p)$.

$$\frac{dl}{dp} = \frac{r}{p} - \frac{n-r}{1-p} = 0 \text{ when } p = \frac{r}{n}.$$

Ex.2 p.91 Let N be the number of tosses up to and including the first heads.

$P(N = r) = (1-p)^{r-1}p$

likelihood $= \{(1-p)^{2-1}p\}\{(1-p)^{5-1}p\} \cdots \{(1-p)^{12-1}p\}$
$= p^5(1-p)^{1+4+2+3+11} = p^5(1-p)^{21}$

which is maximised when $p = \dfrac{5}{26}$.

Ex.3 p.91 Let q be the probability of a packet being returned.
By independence, $q = p^2$.
Likelihood of 25 packets returned $= (p^2)^{25}(1-p^2)^{75} = p^{50}(1-p^2)^{75}$.
Log likelihood $= 50 \log p + 75 \log(1-p^2)$.

$\dfrac{dl}{dp} = \dfrac{50}{p} - \dfrac{150p}{1-p^2} = 0$ when $50 - 50p^2 = 150p^2$, i.e. $p = \dfrac{1}{2}$.

When r out of n are returned, $\hat{p} = \sqrt{\dfrac{r}{n}}$.

95% confidence interval for q (based on 25 successes in 100 trials): $(0.1712, 0.3484)$.
So 95% confidence interval for $p = \sqrt{q}$ is $(0.4137, 0.5902)$.

Ex.4 p.91 Probability of box being returned, $q = 2p - p^2$.
Likelihood $= \{(2p-p^2)\}^{25}\{(1-p)^2\}^{75} = p^{25}(2-p)^{25}(1-p)^{150}$.
Log likelihood $= 25\log p + 25\log(2-p) + 150\log(1-p)$.

$$\frac{dl}{dp} = \frac{25}{p} - \frac{25}{2-p} - \frac{150}{1-p}$$

$= 0$ when $\dfrac{25(2-p)(1-p) - 25p(1-p) - 150p(2-p)}{p(2-p)(1-p)} = 0$.

i.e. $p^2 - 2p + 0.25 = 0$, i.e. $p = 0.13397$.

Ex.5 p.91 Likelihood $= (1-e^{-20\mu})^{10}(e^{-20\mu} - e^{-40\mu})^9(e^{-40\mu} - e^{-60\mu})^7 \cdots (e^{-100\mu})^{13}$
Put $p = e^{-20\mu}$
$= (1-p)^{10}[p(1-p)]^9[p^2(1-p)]^7[p^3(1-p)]^2[p^4(1-p)]^1(p^5)^{13}$
$= (1-p)^{10+9+7+2+1}p^{9+14+6+4+65} = (1-p)^{29}p^{98}$.

So $\hat{p} = \dfrac{98}{98+29} = \dfrac{98}{127}$ giving $\hat{\mu} = 0.01296$.

Expected frequencies: $42 \times (1-\hat{p}) = 9.59$, $42 \times \hat{p}(1-\hat{p}) = 7.4$, etc.

Ex.1 p.93 Let N be the number of calls received in 10 seconds.
$P(N = r) = e^{-10\mu}(10\mu)^r/r!$
So $P(N \geq 1) = 1 - e^{-10\mu} = 0.025$ when $e^{-10\mu} = 0.975$ or $\mu = 0.002532$.
$P(N \leq 1) = e^{-10\mu} + 10\mu e^{-10\mu} = 0.025$ when $\mu = 0.5572$.
95% confidence interval for μ is $(0.002532, 0.5572)$.

Ex.2 p.93 Let N be the number of calls received in 1 minute.
$P(N = r) = e^{-60\mu}(60\mu)^r/r!$

$$P(N \geq 52) = P(N > 51.5) = P\left(\frac{N - 60\mu}{\sqrt{60\mu}} > \frac{51.5 - 60\mu}{\sqrt{60\mu}}\right)$$

$= 0.025$ if $\dfrac{51.5 - 60\mu}{\sqrt{60\mu}} = 1.96$, i.e. $\mu = 0.6537$.

Similarly, $P(N \leq 52) = P(N < 52.5) = 0.025$ if $\dfrac{52.5 - 60\mu}{\sqrt{60\mu}} = -1.96$,

giving $\mu = 1.1459$.
95% confidence interval for μ is $(0.6537, 1.1459)$.

Ex.3 p.94 Let T be a typical time, with probability density $f(t) = \dfrac{1}{\mu}e^{-t/\mu}$.

$P(T > 120) = e^{-120/\mu}$.
Likelihood $= f(21.7)f(107.0)f(24.5)\cdots f(119.9) \times e^{-120/\mu}$.

Log-likelihood
$= -9\log\mu - (21.7 + 107.0 + 24.5 + \cdots + 119.9 + 120)/\mu$.
$= -9\log\mu - 684.85/\mu$.

So $\dfrac{dl}{d\mu} = \dfrac{-9}{\mu} + \dfrac{684.85}{\mu^2} = 0$ when $\hat{\mu} = \dfrac{684.85}{9} = 76.0944$.

If censored value is 240.1, $\hat{\mu} = \dfrac{804.95}{10} = 80.495$.

95% confidence interval: $(49.69, 211.72)$.

170 Statistics

Ex.2 p.96 95% confidence interval for μ is $0.95 \pm 1.96 \times \dfrac{0.14}{\sqrt{50}} = (0.9112, 0.9888)$.

This does not depend on the weights being normally distributed.

The interval just includes 1 if $\dfrac{1.96\sigma}{\sqrt{50}} = 0.05$ or $\sigma = 0.1804$.

Ex.2 p.99 $\sum x_i y_i = 300842$, $\sum x_i^2 = 87762$ so $\hat{b} = \dfrac{300842}{87762} = 3.4279$ miles/mm.

Ex.1 p.101 $E(Y_i) = x_i$, $E(Y_i) = x_i - 5$, $E(Y_i) = 0.95x_i$, $E(Y_i) = 10 + 0.95x_i$.

Ex.2 p.105
| y_i | 2 | 5 | 5 | 20 | $\bar{y} = 8$ | $S_{yy} = 198$ | $S_{xy} = 27$ |
| x_i | 1 | 2 | 3 | 4 | $\bar{x} = 2.5$ | $S_{xx} = 5$ | |

So $\hat{\beta}_1 = \dfrac{27}{5} = 5.4$, $\hat{\beta}_0 = 8 - 5.4 \times 2.5 = -5.5$.

Residuals are 2.1 -0.3 -5.7 3.9 with sum of squares 52.2.

So $\hat{\sigma}^2 = \dfrac{52.2}{2} = 26.1$.

Ex.3 p.105 $\hat{\beta}_1 = \dfrac{120550}{125000} = 0.9644$, $\hat{\beta}_0 = 205.48 - 0.9644 \times 200$.

RSS $= 5645.82$. Therefore $\hat{\sigma} = \sqrt{\dfrac{5645.82}{23}} = 15.667$.

Ex.1 p.106 $\hat{\beta}_1 = \dfrac{S_{xy}}{S_{xx}} = \dfrac{87}{94} = 0.92553$.

$\hat{\beta}_0 = \bar{y} - \hat{\beta}_1 \bar{x} = 9 - 0.92553 \times 8 = 1.5957$.

RSS $= 90 - 0.92553 \times 87 = 9.47872$, so $\hat{\sigma} = \dfrac{9.47872}{8} = 1.0885$.

When $x = 10$, $\hat{y} = 9 + 0.92553 \times (10 - 8) = 10.851$.

Should not use to predict $x = 20$ without further information whether the yield/fertilizer relation is still linear in this region.

Ex.2 p.106 $\hat{b} = \dfrac{\sum y_i x_i^2}{\sum x_i^4} = 0.2535$. Since the observations are on the same bean plant we cannot assume $Y_i = bx_i^2 + E_i$ with E_i independent $N(0, \sigma^2)$. Fitting a straight line would not be adequate.

Ex.3 p.107 $S(a, b) = (16 - a - b)^2 + (11.5 - a)^2 + (4.5 - b)^2 + (5.5 - a + b)^2$ which is maximised when $a = 11$ and $b = 5$.

Chapter 5

Ex.1 p.109 95% confidence interval for μ is $42.3 \pm 1.96 \times 5.5/\sqrt{10} = (38.89, 45.71)$. This does not contain 37.5, so provide tuition.

Ex.1 p.118 $\bar{x} = 55.6$, $s = 3.0258$. $t = \dfrac{55.6 - 55}{3.0258/\sqrt{10}} = 0.627$ (significance probability $= 0.273$).

So the price is not significantly lower. $t_{0.975}(9) = 2.262$.

95% confidence interval: $55.6 \pm 2.262 \times 3.0258/\sqrt{10} = (53.44, 57.76)$.

Ex.2 p.118 $\bar{x} = 100.4$, $s = 3.5071$, $t_{.975}(4) = 2.776$.
95% confidence interval: $100.4 \pm 2.776 \times 3.5071/\sqrt{5} = (96.05, 104.75)$.
Cannot identify the class.

Ex.1 p.120 Mean difference $= 37.6$. Standard deviation $= 40.125$. $t_{.975}(9) = 2.262$.
95% confidence interval: $37.6 \pm 2.262 \times 40.125/\sqrt{10} = (8.9, 66.3)$.

Ex.2 p.120 Mean difference $(B - A) = 0.1875$. Standard deviation $= 0.2295$.
$$t = \frac{0.1875}{0.2295/\sqrt{8}} = 2.31. \text{ Two-tail significance probability} = 0.054.$$

Ex.3 p.120 (a) Mean difference $= 0.6244$. Standard deviation $= 0.9636$.
$$t = \frac{0.6244}{0.9636/\sqrt{118}} = 7.039. \text{ One-tail significance probability} =$$
7×10^{-11}.
(b) Mean difference $= 1.916$. Standard deviation $= 3.341$.
$$t = \frac{1.916}{3.341/\sqrt{118}} = 6.23. \text{ One-tail significance probability} =$$
3.8×10^{-9}.

Ex.1 p.124 $\bar{A} = 3.332$, $s_A = 0.6792$, $\bar{B} = 3.984$, $s_B = 0.3773$, $s = 0.549$,
$t_{.975}(28) = 2.048$. 95% confidence interval:
$$3.984 - 3.332 \pm 2.048 \times 0.549\sqrt{\frac{2}{15}} = (0.241, 1.063) - \text{recommend B}.$$

Ex.3 p.124 $s = \sqrt{\dfrac{2180.4 + 2291.6}{78}} = 7.5719$, $t = \dfrac{2510 - 2503}{7.5719\sqrt{\dfrac{2}{40}}} = 4.134$.

Two-tail significance probability $= 0.00009$. Reject hypothesis of no difference.

Ex.1 p.127 Wilcoxon statistic $= 25$. p-value $= 0.076$. The Wilcoxon test does not reject the null hypothesis at the 5% level.

Ex.3 p.129 $W = 1130$. Significant at 0.0014 (adjusted for ties) and the hypothesis of no difference is rejected. We really should know whether the size of the workforce is essentially the same.

Ex.1 p.132 Double glazing: $\hat{p} = \dfrac{60 + 47}{200} = 0.535$, $\hat{p}_1 - \hat{p}_2 = 0.6 - 0.47 = 0.13$.
Standard deviation of $p_1 - p_2 = \sqrt{0.535 \times 0.465/50} = 0.07054$.
$$P(\hat{p}_1 - \hat{p}_2 > 0.13) = 1 - \Phi\left(\frac{0.13}{0.07054}\right) = 0.033.$$
Alarms: $\hat{p} = \dfrac{35 + 22}{200} = 0.285$, $\hat{p}_1 - \hat{p}_2 = 0.35 - 0.22 = 0.13$.
Standard deviation of $p_1 - p_2 = \sqrt{0.285 \times 0.715/50} = 0.0638$.
$$P(\hat{p}_1 - \hat{p}_2 > 0.13) = 1 - \Phi\left(\frac{0.13}{0.0638}\right) = 0.021.$$
The 5% one-tailed tests used here (because A is thought to be more prosperous than B) both lead to the rejection of the null hypothesis.

Ex.1 p.135 More variation observed on *cloudy* days!
To test this: $s_{cl}^2/s_{sun}^2 = 1.635$, $F_{0.95}(8, 7) = 3.726$. Not significant.

Ex.1 p.139 (a) $Y = 1.4 + 0.857X$, (b) $t = 3.288$, $F = 10.81 = t^2$.

Ex.2 p.141 In each case (almost exactly), $r = 0.816$, $\hat{\beta}_0 = 3$, $\hat{\beta}_1 = 0.5$, $RSS = 13.8$.

Ex.1 p.144 (a) Binomial, (b) $n = 10$, $\hat{p} = 0.1468$,
(c) Frequencies: 102.2 175.9 136.2 62.5 18.8 3.9, 0.6.
(d) Combine 4, 5 and >5: $Q = 2.16$, $\chi^2_{0.95}(4) = 9.49$.

Ex.2 p.149 $Q = 0.8612$, $\chi^2_{0.95}(8) = 15.5$.

Ex.1 p.152 Paired, one-tailed. Mean difference $= 0.773$. $s = 0.579$.

$$t = \frac{0.773}{0.579/\sqrt{10}} = 4.22.\text{ Significance probability} = 0.0011.$$

Ex.2 p.153 Paired t-test. Mean increase in profit $= 185.6$p. $s = 234.712$.

$$t = \frac{185.6}{234.712/\sqrt{15}} = 3.06.\text{ Significance probability} = 0.0042.$$

Ex.3 p.153 (a) Pooled $s = 0.065$, $t = 12.37$ (highly significant).
(b) Pooled $s = 3.325$, $t = 1.464$ (two-tail significance probability $= 0.146$).

Ex.4 p.154 Mann–Whitney, $W = 976.5$. Two-tail significance probability $= 0.000065$.

Ex.6 p.154 For $\beta_0 = 0$, $t = 1.34$ (significance probability $= 0.19$).

$$\text{For } \beta_1 = 1, t = \frac{0.9644 - 1}{\sqrt{245.47043/125000}} = -0.803.\text{ Accept } \beta_0 = 0, \beta_1 = 1.$$

Index

accessibility sampling, 3
arithmetic mean, 20–2
averages, 19–25
  arithmetic mean, 20–2
  median, 22–3
  percentiles, 24–5

Bernoulli trials, 42, 85
bias, 84
binomial distribution, 42–8, 57
  negative, 56–7, 58
  and normal distribution, 66–9
  and Poisson distribution, 51–3
  proportions, testing of, 130–2
boxplots, 25, 27–8

central limit theorem, 74–5, 79
charts, 4
  pie, 13–14
chi-squared distribution in hypothesis testing, 132
chi-squared goodness-of-fit tests in hypothesis testing, 142–50, 152
  contingency tables, 146–9, 152
  too good fit, 145–6
classes, in histograms
  boundaries, 11
  choice of, 12
  interval classes, 11
  limits, 11
  representative, 11
classificatory scale, 5
cluster sampling, 3
confidence intervals
  in estimation, 86
    formula for, 87–8
  in hypothesis testing, 108–9
  in regression and correlation, 138–9
  sample variance distribution, 132–4
  t distribution, 117
confirmatory analysis, 10
contingency tables in chi-squared goodness-of-fit tests, 146–9
continuous distributions, 58–60
  mean and variance of, 60–1
correlation, 32–5
  coefficients in, 152
  hypothesis testing in, 135–42
    confidence intervals, 138–9
    linear coefficient, 140–1
    linear relationships, 136–7
    variance, estimation, 137–8
cumulative distribution functions, 59, 60

data
  analysis, 4–5
  collection, 1–4
  dredging, 8
  experimental, 2
  observational, 2
  primary, 1
  secondary, 2
  sets, 157–61
decision rule in hypothesis testing, 110
deduction, 4–5
degrees of freedom in hypothesis testing, 116–17
descriptive statistics, 4, 82
difference of means in hypothesis testing, 120–5, 150–1
  unequal variance, 125
  unknown variance, 122–4
discrete data in hypothesis testing, 113–14
discrete random variables, 40
dispersion, 25–7

error types in hypothesis testing, 110–12
estimation, 82–107
  least squares method in, 95, 97–100
  linear regression in, 100–5
  maximum likelihood *see* maximum likelihood estimation
  simple problem in, 83–4
expected values, 40–2
  in normal distribution, 70–2
experimental data, 2
explanatory variable in linear regression, 36
exploratory data analysis (EDA), 4, 10
  averages, 19–25
    arithmetic mean, 20–2
    median, 22–3
    percentiles, 24–5
  boxplots, 25, 27–8
  correlation, 32–5
  dispersion, 25–7
  frequency distributions, 10–11
  histograms, 11–13
  multiple displays, 17–19
  paired values, 28–32

exploratory data analysis (EDA) *cont.*
  regression analysis, 35–9
  stem-and-leaf displays in, 14–17
exponential distribution, 61–3, 79
  maximum likelihood estimation in, 88–91, 93–4

F distributions in hypothesis testing, 134
finite population, sampling from, 162–3
five-number summary, 25
frequency distributions, 10–11
frequency tables in maximum likelihood estimation, 88–91

general normal distribution, 65–6
geometric distribution, 55–6, 58
graphs, 4

hinges, 24
histograms, 11–13
  multiple displays, 17–19
hypergeometric distribution, 54–5, 58
  binomial approximation to, 55
hypothesis testing, 6, 108–54
  chi-squared tests, 142–50
    contingency tables, 146–9
    too good fit, 145–6
  classical formulation, 109–13
    decision rule, 110
    error types, 110–12
    null and alternative, 109–10
    one-tailed and two-tailed tests, 112–13
  confidence interval approach, 108–9
  difference of means, 120–5, 150–1
    unequal variance, 125
    unknown variance, 122–4
  discrete data, 113–14
  matched pairs, 118–20, 150
  non-parametric tests, 125–30
    Mann–Whitney, 127–9, 151
    Wilcoxon signed-ranks, 125–7, 151
  normal distribution, unknown variance, 115–18
    degrees of freedom, 116–17
    Student's t distribution, 116
    Student's t-test, 117–18, 150
    t distribution, confidence intervals, 117
  on proportions, 130–2, 151
    differences between, 131
  regression and correlation, 135–42
    confidence intervals, 138–9
    linear coefficient, 140–1
    linear relationships, 136–7
    variance, estimation, 137–8
  sample variance distribution, 132–5
    chi-squared distribution, 132
    confidence intervals, 132–4
    F distribution, 134
    variance-ratio test, 134–5, 151
  significance probability, 114–15
  simple and complex, 109

induction, 4–5
inferential statistics, 4, 82
interquartile range, 27
interval estimates, 108
interval scale, 5
interviews, 4

judgemental sampling, 3

least squares method, 36
  in linear regression, 102–3
  in maximum likelihood estimation, 95, 97–100, 106
likelihood functions in estimation, 85
linear interpolation, 24
linear models, generalized, 155
linear regression
  by computer, 105
  in estimation, 100–5
  least squares in, 102–3
  simple, 36–7
  variables in, 36
log-likelihood functions in estimation, 90–1, 106

mail surveys, 4
Mann–Whitney hypothesis tests, 127–9, 151
matched pairs, hypothesis testing in, 118–20, 150
maximum likelihood estimation, 85–8, 106
  for exponential distributions, 88–91, 93–4
  frequency tables in, 88–91
  for normal distribution, 94–7
  for Poisson distributions, 92–3
  for probability distributions, 92–4
means
  arithmetic, 20–2
  of continuous distributions, 60–1, 79
  as expected value, 41
  in normal distribution, 70–2
measurement, scales, 5–6
median, 22–3
modes, 23
multiple displays, 17–19
multivariate statistical analysis, 155

negative binomial distribution, 56–7, 58
nominal scale, 5
non-parametric tests, 125–30
normal distribution, 63–78
  binomial distribution, approximation to, 66–9

normal distribution *cont.*
  central limit theorem, 74–5, 79
  expected values, 70–2
  general, 65–6
  hypothesis testing, unknown variance
    degrees of freedom, 116–17
    Student's t distribution, 116
    Student's t-test, 117–18, 150
    t distribution, confidence intervals, 117
  linear combinations of random variables, 72–4
  maximum likelihood estimation in, 94–7
  Poisson distribution, approximation to, 69
  simulation in, 75–7
  standard, 63–5
null hypothesis, 109–10

observational data, 2
one-tailed hypothesis tests, 112–13
  significance probability in, 115
ordinal scale, 5

paired values, 28–32
  scatter plots, 29–32
parameter, estimation of, 84, 106
Pearson correlation coefficient, 32–5
percentiles, 24–5
personal interviews, 4
pie charts, 13–14
point estimates, 108
Poisson distribution, 48–54, 57–8
  approximation to binomial distribution, 51–3
  maximum likelihood estimation for, 92–3
  normal distribution, approximation to, 69
population, 2
  linear correlation coefficient, 140
primary data, 1
probability, 40–2
  density functions, 59, 78
    in maximum likelihood estimation, 89
  distributions, maximum likelihood estimation for, 92–4
  significance, in hypothesis testing, 114–15
proportions, testing of, 130–2, 151
  differences between, 131

quality control, 7
quartiles, 24
  deviation, 27
  and hinges, 24–5
  spread measured, 27

random sample, 3
random variables, 40–2
  dependent, 163

range, 26
ranking scale, 5
ratio scale, 5
regression analysis, 35–9
  coefficients in, 152
  hypothesis testing in, 135–42
    confidence intervals, 138–9
    linear coefficient, 140–1
    linear relationships, 136–7
    variance, estimation, 137–8
  simple linear regression, 36–7
representative sample, 3
residual sum of squares, 102–3, 137–9
response variable in linear regression, 36

sample fraction, 163
sample regression coefficients, 37
sample statistics
  averages, 19–20
  in sampling distributions, 78
sample variance distribution, 132–5
  chi-squared distribution, 132
  confidence intervals, 132–4
  F distribution, 134
  variance-ratio test, 134–5, 151
sampling, 2–4
  accessibility, 3
  cluster, 3
  distributions, 78
    of estimator, 84, 106
  from finite population, 162–3
  judgemental, 3
  with replacement, 162
  surveys, 4
  systematic, 3
  without replacement, 162–3
scales, 5–6
scatter plots, 29–32
SD line, 35–6
selective reporting, 8
semi-interquartile range, 27
significance level of hypothesis test, 111
significance probability in hypothesis testing, 114–15
simple linear regression, 36–7
simulation in normal distribution, 75–7
single samples: maximum likelihood estimation for, 94–7
skewness, 20
standard deviation, 26
standard normal distribution, 63–5
statistics
  abuses of, 7–8
  applications of, 6–7
stem-and-leaf displays, 14–17
  median in, 22

stratified random sample, 3
Student's *t* distribution in hypothesis testing, 116
Student's *t*-test in hypothesis testing, 117–18, 150
sufficient statistic in Poisson distribution, 92
sum of squares in maximum likelihood estimation, 95
   in linear regression, 102–3
   within-groups sums, 98
systematic sampling, 3

t distribution in hypothesis testing, 117
telephone interviews, 4
two-tailed hypothesis tests, 112–13

unbiasedness, 84

variance, 26
   analysis of, 155
   of continuous distributions, 60–1, 79
   of estimators, 84
   of expected values, 41
   in hypothesis testing
      in correlation, 137–8
      unequal, 125
      unknown
         in difference of means, 122–4
         in normal distributions, 117–18
   in least squares method of estimation, 98
   in normal distribution, 70–2
variance-ratio tests in hypothesis testing, 134–5, 151

Wilcoxon signed-ranks hypothesis tests, 125–7, 151